Reliability Models of Complex Systems for Robots and Automation

Reliability Models of Complex Systems for Robots and Automation

Hamed Fazlollahtabar
Department of Industrial Engineering
Sharif University of Technology
and
Iran National Elites Foundation
Tehran, Iran

and

Seyed Taghi Akhavan Niaki
Department of Industrial Engineering
Sharif University of Technology
Tehran, Iran

CRC Press
Taylor & Francis Group
Boca Raton London New York

CRC Press is an imprint of the
Taylor & Francis Group, an **informa** business

CRC Press
Taylor & Francis Group
6000 Broken Sound Parkway NW, Suite 300
Boca Raton, FL 33487-2742

© 2017 by Taylor & Francis Group, LLC
CRC Press is an imprint of Taylor & Francis Group, an Informa business

No claim to original U.S. Government works

Printed on acid-free paper

International Standard Book Number-13: 978-1-138-56963-8 (Paperback)
International Standard Book Number-13: 978-1-138-56966-9 (Hardback)

Visit the Taylor & Francis Web site at
http://www.taylorandfrancis.com

and the CRC Press Web site at
http://www.crcpress.com

Contents

Preface

The word *reliability* can be traced back to 1816, when it was used by the poet Coleridge. Before World War II (WWII), the word was linked mostly to repeatability (Saleh and Marais, 2006). A test (in any science) was considered reliable if the same results were obtained repeatedly. In the 1920s, Dr. Walter A. Shewart promoted product improvement through the use of statistical process control at Bell Labs. Around this time, Wallodi Weibull was working on statistical models for fatigue. The development of reliability engineering was on a parallel path with quality (Juran and Gryna, 1988). The modern definition of the word *reliability* was given by the U.S. military in the 1940s and has evolved to the present usage. It initially came to mean a product would operate when expected (nowadays called "mission readiness") and for a specified period. During WWII and sometime later, many reliability issues were caused due to defects in electronic components and fatigue. In 1945, M.A. Miner published a seminal paper titled "Cumulative Damage in Fatigue" in an ASME journal. The main application of reliability engineering in the military was for vacuum tubes used in radar systems and other electronics, for which reliability had proven very problematic and costly. The IEEE formed the Reliability Society in 1948. In 1950, on the military side, a group called the Advisory Group on the Reliability of Electronic Equipment, AGREE, was born. This group recommended the following three main ways of working:

1. Improve component reliability
2. Establish quality and reliability requirements (also) for suppliers
3. Collect field data and find root causes of failures

In the 1960s, emphasis was given to reliability testing at the component and system levels. The famous military standard 781 was created at that time. Also, around this period, the much-used (and also much-debated) military handbook 217 was published by Radio Corporation of America (RCA) and was used for the prediction of failure rates of components. The emphasis on component reliability and empirical research (e.g., Mil Std 217) alone slowly decreased. More pragmatic approaches, as used in the consumer industries, began to be used. The 1980s was a decade of great changes. Televisions had become semiconductor based. Automobiles rapidly increased their use of semiconductors with a variety of microcomputers under the hood and in the dashboard. Electronic controllers were developed for large air conditioning systems, as for microwave ovens and a variety of other appliances. Communications systems began to adopt electronics to replace older mechanical switching systems. Bellcore issued the first consumer prediction methodology for telecommunications, and SAE developed a similar document, SAE870050, for automotive applications. The nature of predictions evolved during the decade, and it became apparent that die complexity was not the only factor that determined failure rates for integrated circuits (ICs). Kam Wong published a chapter questioning the bathtub curve (Wong, 1981; see also O'Connor, 2002). During this decade, the failure rate of many components dropped by a factor of 10. Software became important to the reliability of systems. By the 1990s, the pace of IC development was picking up. Advanced use of stand-alone microcomputers was common, and the PC market helped to keep IC densities following Moore's law and doubling about every 18 months. Reliability engineering was shifting more toward understanding the physics of failure. Failure rates for components kept dropping, but system-level issues became more prominent. Product development time continued to shorten throughout this decade, and what had been done in 3 years was being done in 18 months. This meant that reliability tools and tasks must be more closely tied to the development process itself. In many ways, reliability became part of everyday life and consumer expectations.

This book considers a complex production system composed of several autonomous robots performing manufacturing jobs. Due to the high technology in robotic systems, their availability became significant. The reliability of such complex systems is analyzed in different conditions, including different objectives. Therefore, various reliability models are developed, each focusing on the specifications of a unique problem and providing reliability evaluation of complex systems. Reliability assessment of robotic-based complex systems has not been investigated and thus is a novel contribution of this book.

Key Terms and Definitions

Binary system:	A system in which information can be expressed by combinations of the digits 0 and 1, or a system consisting of two parts.
Complex system:	A system having multiple components and subcomponents, making it difficult to evaluate its performance.
Failure:	Failure is the state or condition of not meeting a desirable or intended objective and may be viewed as the opposite of success.
Reliability:	Reliability is the ability of a system or component to perform its required functions under stated conditions for a specified period of time.

Repair: To restore to sound condition after damage or injury.

Robot: An automated driverless controllable vehicle, which is used as a transportation and transferring device.

Work station: An industrial unit for processing a function allocated as per the manufacturing plan using input material and delivering semiproduced or final products to the next manufacturing unit.

Acknowledgments

We express our gratitude to the many people who saw us through this book, and to all those who provided support, talked things over, read, wrote, offered comments, allowed us to quote their remarks, and assisted in the editing, proofreading, and design of this book.

We thank Iran National Elites Foundation and Sharif University of Technology for enabling us to publish this book. Above all, we thank our family, who supported and encouraged us in spite of all the time it took us away from them. It was a long and difficult journey for them.

Last but not least, we beg forgiveness of all those who have been with us over the course of the years and whose names we have failed to mention.

About the Authors

Hamed Fazlollahtabar earned a BSc and an MSc in industrial engineering from Mazandaran University of Science and Technology, Babol, Iran, in 2008 and 2010, respectively. He received his PhD in industrial and systems engineering from Iran University of Science and Technology, Tehran, Iran, in 2015. He recently completed a postdoctoral research fellowship at Sharif University of Technology, Tehran, Iran, in the area of reliability engineering for complex systems. He is on the editorial boards of several journals and on the technical committees of several conferences. His research interests are in robot path planning, reliability engineering, supply chain planning, and business intelligence and analytics. He has published more than 230 research papers in international books, journals, and conferences. He has also published five books out of which three are internationally distributed to academicians.

Seyed Taghi Akhavan Niaki is professor of industrial engineering at Sharif University of Technology, Tehran, Iran. His research interests are in the areas of simulation modeling and analysis, applied statistics, multivariate quality control, and operations research. Before joining Sharif University of Technology, he worked as a systems engineer and quality control manager for Iranian Electric Meters Company. He received his BSc in industrial engineering from Sharif University of Technology in 1979, and his master's and PhD, both in industrial engineering, from West Virginia University in 1989 and 1992, respectively. He is the co-editor-in-chief of *Scientia Iranica, Transaction E*, a board member of several international journals, and a member of Alpha Pi Mu.

1

Introduction and Background

1.1 Overview

Today's competitive world with increasing customer demands for highly reliable products makes reliability engineering a challenging task. Reliability analysis is one of the main tools to ensure agreed delivery deadlines are met, which in turn maintains certainty intangible factors such as customer goodwill and company reputation (Jardine, 1998). Downtime often leads to both tangible and intangible losses. These losses may be due to some unreliable components; thus, an effective strategy needs to be framed for maintenance, replacement, and design changes needed for those components (Ross, 1970; Billinton and Allan, 1983; O'Connor, 2002).

The design for reliability is an important research area, specifically in the early design phase of product development. In fact, reliability should be designed and built into products and systems at their earliest development stages. Reliability-targeted design is the most economical approach to minimize the life cycle costs of the product or system, based on which one can achieve better product or system reliability at much lower costs. Otherwise, the majority of life cycle costs are locked in phases other than design and development; poor reliability consideration at the design stage can have implications later on in the product life. If reliability analysis is applied during the conceptual design phase, its impact will be more remarkable on the design process, producing high-quality products (Soleimani and Pourgol-Mohammad, 2014). In other words, a structure reliable in concept is less expensive than a structure that is not reliable in concept, even with improvement at a later phase of the design process (Avontuur and van der Werff, 2001). Moreover, reliability analysis in the conceptual design process leads to more optimal structures than its application at the end of the design process (Avontuur, 2000).

In problems of maintenance optimization, it is convenient to assume that repairs are equivalent to replacements, and those systems or objects are, therefore, brought back to an as-good-as-new state after each repair. Standard results in renewal theory may then be applied for determining optimal maintenance policies. In practice, there are many situations in which this assumption cannot be made. The quintessential problem with imperfect maintenance is how to model it. In many cases, it is very difficult to assess by how much a partial repair will improve the condition of a system or object, and it is equally difficult to assess how such a repair influences the rate of deterioration. Kallen (2011) proposed a superposition of the renewal process that is used to model the effect of imperfect maintenance. It constituted a different modeling approach than the more common use of a virtual age process.

Nishijima (2007) addressed the issue of optimization of reliability acceptance criteria for components of complex engineered systems with a given criterion for an acceptable system risk. To this end, they first described how complex engineered systems may be modeled hierarchically using Bayesian probabilistic networks. The Bayesian probabilistic network serves as a function that relates the reliability acceptance criteria of the individual components of the system to the risk acceptance criteria of the system. Thereafter, a constrained optimization problem is formulated for the optimization of the component reliabilities. In this optimization problem, the system risk acceptance criteria define the constraint, and the expected utility from the system is considered the objective function.

During the design phase of a product, reliability engineers are called upon to evaluate the reliability of the system. The question of how to meet a reliability goal for the system arises when the estimated reliability is inadequate. This then becomes a reliability allocation problem at the component level. Mettas (2000) estimated a general model for the minimum reliability requirement of multiple components within a system that will yield to the ultimate reliability of the system. The model consisted of two parts. The first part was a nonlinear programming formulation of the allocation problem. The second part was a cost function formulation to be used in the nonlinear programming algorithm, where a general behavior of the cost as a function of a component's reliability was assumed.

In the exact method, there are two classes for the computation of network reliability. The first class deals with the enumeration of all the minimum paths or cuts. A path is a subset of components (edges and/or vertices) that guarantees the source and the sink to be connected if all the components of this subset are functioning. A path is minimal if a subset of elements in the path, which is also a path, does not exist. A cut is a subset of components (edges and/or vertices) whose failure disconnects the source and the sink. A cut is minimal if a subset of elements in the cut, which is also a cut, does not exist (Hariri and Raghavendra, 1987; Ahmad, 1988). The probabilistic evaluation uses the inclusion–exclusion or sum-of-disjoint-products methods because this enumeration provides nondisjoint events. Numerous works about these kinds of methods have been presented in the literature (Lucet and Manouvrier, 1999). In the second class, the algorithms are based on graph topology. In the first process, we reduce the size of the graph by removing some structures, namely, polygon-to-chain (Choi and Jun, 1985) and delta-to-star reductions (Gadani, 1981). By this, we will be able to compute the reliability in linear time and the reduction will result in a single edge. The idea is to decompose the problem into one failed and another functioning (Carlier and Lucet, 1996). The same was confirmed by Theologou and Carlier (1991) for dense networks. Satyanarayana and Chang (1983) and Wood (1985) have shown that factoring algorithms with reductions are more efficient at solving this problem than the classical path or cut enumeration methods.

Statistical modeling has been paramount for the quality control and maintenance of repairable production systems. It allows to reduce costs and to prevent the occurrence of undesirable events. At the same time, it provides support for enhancing production levels and the longevity of components. The generalized renewal process (GRP) is a powerful statistical formalism for modeling repairable systems. It enables one to evaluate the quality of the performed interventions as well as to forecast the time for undesirable events to occur (Felix de Oliveira et al., 2016).

In GRP reliability analysis for repairable systems, the Monte Carlo (MC) simulation method is often used, instead of the numerical method, to estimate model parameters because of the complexity and difficulty of developing a mathematically tractable probabilistic model. Wang and Yang (2012) proposed a nonlinear programming formulation based on the conditional Weibull distribution for repairable systems, using negative log-likelihood as an objective function and adding inequality constraints to model parameters to estimate the restoration factor for the Kijima-type GRP model. This method minimized the negative log-likelihood directly and avoided solving the complex system of equations. Numerical studies on NC machine tools were analyzed by the proposed approach. The results showed that the GRP model is superior to the ordinary renewal process (ORP).

An important characteristic of the GRP, which is of great practical interest, is the generalized renewal equation, which represents the expected cumulative number of recurrent events as a function of time. Just like in an ORP, the problem is that the generalized renewal equation does not have a closed form solution, unless the underlying event times are exponentially distributed. The MC solution, although exhaustive, is computationally demanding. Yevkin and Krivtsov (2012) offered a simple-to-implement (in an Excel spreadsheet) approximate solution, when the underlying failure-time distribution is Weibull. The accuracy of the proposed solution was in the neighborhood of 2%, when compared to the respective MC solution.

Many models and methodologies are available to deal with imperfect repair for repairable systems. Initial attempts at modeling imperfect repair using the (p, q) rule that defined the two extremities of imperfect repair—perfect renewal and minimal repair—were effectively extended by Kijima and Sumita. They developed a generalized renewal theory from the renewal theory in the context of imperfect repair and applied it to repairable systems with the concept of virtual age. Since this pioneering work, much of imperfect repair modeling literature builds up on Kijima's models based on the GRP. Tanwar et al. (2014) conducted a survey for imperfect repair of repairable systems using the GRP based on arithmetic reduction of age (ARA) and arithmetic reduction of intensity (ARI) concepts in general and Kijima models in particular. In addition to the theoretical development of Kijima models and its extensions, the review highlighted their applications such as designing maintenance policies based on the concept of ARA.

van der Weide and Pandey (2015) presented a stochastic approach to analyze instantaneous unavailability of standby safety equipment caused by latent failures. The problem of unavailability analysis was formulated as a stochastic alternating renewal process without any restrictions on the form of the probability distribution assigned to time to failure and repair duration. An integral equation for point unavailability was derived and numerically solved for a given maintenance policy.

Alem Tabriz et al. (2015) considered both failure rates as internal factors and the shocks as an external factor to develop age-based replacement models in order to determine the optimal replacement cycle. As a result, according to system reliability, maintenance costs of the system were minimized. Analysis of results showed all models provided optimal replacement cycle, and at this time, the cost rate of the system, by considering the reliability rate, is minimal. Also, with an increase of one to two units, the reliability rate increases much higher than the cost.

1.2 Reliability Engineering

Among its many interpretations, the term reliability most commonly refers to the ability of a device or system to perform a task successfully when required. More formally, it is described as the probability of proper functioning at a given time and under specified operating conditions. Mathematically, the reliability function is defined by

$$R(t) = P(T > t); \quad \forall t \geq 0, \tag{1.1}$$

where T is a nonnegative random variable representing the device or system lifetime. For a system composed of at least two components, the system reliability is determined by the reliability of the individual components and the relationships among them. These relationships can be depicted using a reliability block diagram (RBD). While simple systems are usually represented by RBDs with components in either a series or a parallel configuration, the RBDs of complex systems cannot be reduced to series, parallel, or series–parallel configurations. In a series system, all components must function satisfactorily in order for the system to operate. For a parallel system to operate, at least one component must function correctly.

There are some techniques available in the literature that can be employed to determine a mathematical expression denoting the reliability of a system in terms of the reliabilities of its components. In these techniques, it is assumed that the reliabilities of the components have been determined using standard (or accelerated) life data analysis techniques, so that the reliability function for each component is known. Having this component-level reliability information available, it then becomes necessary to determine how these component reliability values are combined to determine the reliability function of the overall system.

Reliability assessment is a systematic implementation through the design, testing, production, storage, and usage phases of a product or system. It is a process of analyzing and confirming the reliability of a system and its components (Hwang et al., 1981; Zio and Pedroni, 2009; Levitin and Lisnianski, 2013). Reliability evaluation includes both qualitative and quantitative analytical techniques to model and predict system reliability throughout the product/system life cycle. There are basically four aspects of the technical contents of system reliability assessment, including reliability modeling, reliability data collection and processing, unit reliability assessment, and system reliability synthesis. To obtain the reliability of a complex system, the reliability model should be built to describe the failure logic relationship between the whole system and its compositions. In recent decades, various reliability modeling methods have been developed for complex systems (Mi et al., 2016), where some static and dynamic modeling techniques, such as RBD model, fault tree (FT) model, binary decision diagrams (BDD) model, Markov model (Li et al., 2012), dynamic fault tree (DFT) model (Li et al., 2013), and Petri net model (Huang et al., 2010), have been applied.

The DFT model, first proposed by Dugan et al. (1992, 2000), is a mature and important method in the reliability analysis of dynamic systems (Hao et al., 2014). In this regard, Rao et al. (2009) presented an approach to solve dynamic gates, which can be used to alleviate the state space explosion problem. Considering the interactive repeated events in different dynamic gates, Merle et al. (2010) developed a new analytical method to solve DFTs with priority dynamic gates and repeated events. In addition, approximate DFT calculations were presented by Lindhe et al. (2012) based on a Markovian approach, which was used for water supply risk modeling performed by standard MC simulations. To overcome the limitations caused by the increasing size of FTs in traditional reliability assessment, Chiacchio et al. (2013) proposed a Weibull-based composition approach for large DFTs to reduce the computational effort. Mo (2014) developed a multivalue, decision-diagram-based DFT analysis method to analyze the reliability of large dynamic systems based on the state explosion and computational efficiency problems in the DFT model.

Moreover, Ge et al. (2015) presented an improved sequential BDD method for quantitative analysis of DFTs with interactive repeated events. It should be noted that in view of the dynamic characteristics in modern complex systems and taking advantage of the dynamic modeling ability of DFTs, a DFT model should be built on the basis of system structure and failure behaviors (Mi et al., 2016).

During the design phase of a product, reliability engineers are called upon to evaluate the reliability of a system. The question of how to meet a reliability goal for the system arises when the estimated reliability is inadequate. This then becomes a reliability allocation problem at the component level. Mettas (2000) developed a general model to estimate the minimum reliability requirement for multiple components within a system that will yield to the ultimate reliability of the system. The model consisted of two parts. The first part was a nonlinear programming formulation of the allocation problem. The second part was a cost function formulation to be used in the nonlinear programming algorithm, where a general behavior of the cost as a function of a component's reliability was assumed. The system cost was then minimized by solving for an optimum component reliability, which satisfied the system reliability goal requirement.

Nishijima (2007) addressed the issue of optimization of reliability acceptance criteria for components of complex engineered systems with a given criterion for an acceptable system risk. They first described how complex engineered systems may be modeled hierarchically using Bayesian probabilistic networks. The Bayesian probabilistic network serves as a function that relates the reliability acceptance criteria of the individual components of the system to the risk acceptance criteria of the system. Thereafter, a constrained optimization problem was formulated for the optimization of the component reliabilities. In this optimization problem, the system risk acceptance criteria defined the constraint, and the expected utility of the system was considered as the objective function. A ship hull structure was taken as an example of a complex engineered system to illustrate how the proposed framework might be implemented into a software tool using commonly available techniques and algorithms.

In maintenance optimization problems, it is convenient to assume that repairs are equivalent to replacements, and those systems or objects are, therefore, brought back to an as-good-as-new state after each repair. Standard results in renewal theory may then be applied for determining optimal maintenance policies. In practice, there are many situations in which this assumption cannot be made. The quintessential problem with imperfect maintenance is how to model it. In many cases, it is very difficult to assess by how much a partial repair will improve the condition of a system or object, and it is equally difficult to assess how such a repair influences the rate of deterioration. Kallen (2011) proposed a superposition of the renewal process to model the effect of imperfect maintenance. It constituted a different modeling approach than the more common use of a virtual age process.

In the past few years, a considerable amount of work has been devoted to improve the efficiency of methodologies applied to reliability and safety analyses of industrial plants. In particular, the need for a more detailed analysis of the system under study is growing, whereby the plant structure as well as its working conditions are taken into account. This implies the evaluation of the interaction of several elements, namely, physical transient, control system intervention, and operator tasks, which are very important during operational or abnormal situations (Apostolakis et al., 1984). Much effort has been devoted to fill the gap between deterministic dynamic analyses of plants, that is, engineering simulations particularly useful to study small configurations, and classical reliability methods, such as logical methodologies like FTs and event trees (ETs), necessary to study complex systems as a whole (Vesely and Goldberg, 1977). New approaches have been embarked on: As an example, Jeong et al. (1987) introduced the Markov chain in the FT analysis to permit the evaluation of the probabilistic behavior of system unavailability versus time, when the plant is decomposed into a reasonable number of supercomponents; other studies have been developed to assess system reliability through a dynamic and qualitative approach based on the Petri nets theory and Markov chains (Bobbio, 1988; Leroy, 1989). The GO FLOW method has been realized to enhance the GO characteristics with the introduction of a time-ordered analysis (Williams and Gateley, 1978; Matsuoka and Kobayashi, 1988). All these methodologies show very interesting approaches to system analysis, but they present two main problems:

1. The effort needed for system decomposition into supercomponents or the construction of charts
2. The adequacy of the process model used in the probabilistic analysis of the system to represent the actual interaction between the process physics and component behavior

The first issue is related to the need to study the reliability of a complex system in detail, and it does not seem to be avoidable. On the contrary, the second problem can be tackled by a more accurate study of the physical and dynamic behavior of the system; a number of methods are being developed with particular attention to this problem. As an example, the approach proposed by Hassan and Aldemir (1990) separates the physical and probabilistic analyses and realizes a suitable method to assess top event sensitivity to uncertainty on the component failure data. Other methods are the Markov failure modeling (Aldemir, 1987) and the continuous event tree method (Smidts, 1990), which link the system model to probabilistic treatment using Markovian or semi-Markovian chains in a complete theoretical analysis, but seem to be very difficult to apply on large configurations; the DYLAM attainment and the possible recovery of "top" methodology, which uses a quantitative dynamic process model for probabilistic analysis; and finally the dynamic event tree method (Siu and Acosta, 1991), which is similar to the DYLAM approach, but deals with the system at a lower level of detail in order to reduce the computation efforts.

There is extensive literature on reliability characteristics of repairable systems with two or three components under varying assumptions on the failures and repairs. In most of these chapters, exponential distributions are assumed for mathematical convenience. The concept of reliability can also be applied in other fields using different techniques. Mahajan and Singh (1999) discussed the reliability analysis of a utensils-manufacturing plant. Goel and Singh (1995) presented reliability analysis of a standby complex system having imperfect switch-over device and availability analysis of a butter-manufacturing system in a dairy plant. Singh (1989) suggested some applications of reliability technologies such as fertilizer industry, sugar industry, and biogas plant. Dhillon and Natesan (1983) discussed power systems in a fluctuating environment. Dayal and Singh (1992) studied reliability analysis of a systems in a fluctuating environment. Kumar et al. (1988) discussed feeding systems in the sugar industry.

Traditional system reliability analysis methodologies are based on "bottom-up" relationships between system and component reliabilities, such as the methodologies explained by Hoyland and Rausand (2004), Kumamoto and Henley (2000). This approach drives reliability analysis toward understanding component reliability characteristics, which then allows system-level reliability prediction. Recent techniques allow reliability analysis to be conducted at the subsystem or system level (which is referred to as higher level as it appears "higher" in many visualization methodologies). The parameters that describe the reliability characteristics of components define the reliability characteristics of the system. Accordingly, these parameters will always be the unknowns of interest in any system reliability analysis, and, correspondingly, reliability analysis inherently involves the "downward" propagation of information. Conversely, reliability prediction is an "upward" expression of information.

In recent years, the requirement of modern technology, especially the complex systems used in the industry, has led to a growth in the number of researches on the design for reliability. Avontuur and van der Werff (2001) and Avontuur (2000) emphasized the importance of reliability analysis in the conceptual design phase. It is demonstrated that it is possible to improve a design by applying reliability analysis techniques in the conceptual design phase. The aim is to quantify the cost of failure and unavailability and compare them with the investment cost to improve the reliability (Abo Al-Kheer et al., 2011).

Following the increase in using automatic systems, and the problem of performance reliability in such equipment, some indexes such as accessibility, rate of fault, etc., are suggested. Since most automatic systems are designed for continuous missions and the destruction during the mission can result in high expenses for utilizers, evaluating the assurance on equipment must be considered in different steps and also in the planning phase to prevent such unwanted destructions (faults) during the work (Fiorenzo, 2008). In this field, Korayem and Iravani (2008) have promoted the reliability and improvement of robot 3P and robot 6R by using the FMEA and QFD tools.

Structural design via deterministic mathematical programming techniques has been widely accepted as a viable tool for engineering design (Haftka and Gürdal, 1992). However, in most structural engineering applications, response predictions are based on models involving uncertain parameters. This is due to lack of information about the value of system parameters external to the structure, such as environmental loads, or internal, such as system behavior. Under uncertain conditions, the field of reliability-based optimization provides a realistic and rational framework for structural optimization, which explicitly accounts for the uncertainties (Enevoldsen and Sørensen, 1994; Kuschel and Rackwitz, 1997; Royset et al., 2001).

Although risk assessment evaluation of complex dependable systems can be performed through the use of dynamic stochastic modeling, in the real industrial world, the well-known combinatorial techniques, such as RBD and static fault tree (SFT), are still the most widely used ones (Stamatelatos et al., 2002).

1.3 Complex Systems

Generally, a complex system in reliability is defined as a system having a combination of series, parallel, R out of N, and standby components. Each of these models has corresponding mathematical formulations for reliability computations leading to decomposition of the original system (or subsystem) into an equivalent one with a known cumulative distribution function (CDF) or reliability function. Continuing the decomposition procedure enables the decision maker to reduce the whole system to a unique component with a known CDF. For better understanding, an illustrative example for a complex system reduction is given in Figure 1.1. The system is composed of both series and parallel components, which are reduced first to a series system and eventually to a one-component system.

It should be noted that the reduction methods explained before are not effective for all systems. In cases with complicated interrelations of components, it is required to develop an efficient methodology. This methodology deals with subjects such as event trees, Boolean representations, coherent structures, cut sets, and decompositions.

Network reliability analysis receives considerable attention for the design, validation, and maintenance of many real-world systems, such as production, computer, communication, or power networks. The components of a

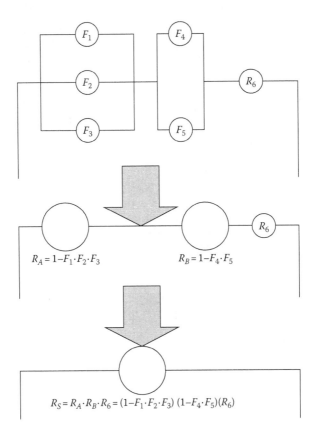

FIGURE 1.1
A complex system reduction.

network are subject to random failures, as more and more enterprises become dependent upon network or networked computing applications. Failure of a single component may directly affect the functioning of a network. So the failure probability of each component is a crucial consideration while considering the reliability of a network. There are so many exact methods for the computation of network reliability (Bobbio et al., 2006). The network model is a directed stochastic graph $G = (V, E)$, where V is the vertex set and E is the set of directed edges. An incidence relation associates with each edge of G a pair of nodes of G, called its end vertices. The edges represent components that can fail with known probability. In real problems, these probabilities are usually computed from statistical data. The problem related with the connection function is NP-hard. The same thing is observed for planar graphs (Provan, 1986).

Complex systems are characterized by large numbers of components, cut sets or link sets, or by statistical dependence between the component states. These measures of complexity render the computation of system reliability a challenging task. Der Kiureghian and Song (2008) developed a decomposition approach that, together with a linear programming formulation, allows determination of bounds on the reliability of complex systems with manageable computational effort. The approach also facilitated multiscale modeling and analysis of a system, whereby varying degrees of detail can be considered in the decomposed system.

Systems can also contain components arranged in both series and parallel configurations. In other words, a complex system is one that cannot be broken down into groups of series and parallel components. In addition, in many cases, it is not easy to recognize which components are in series and which are in parallel. The network shown in Figure 1.2 is a good example of such a complex system.

As Figure 1.2 illustrates, this system cannot be broken down into a group of series and parallel systems. This complicates the problem of determining the system's reliability. If the system can be broken down into series/parallel configurations, it is a relatively simple matter to determine the mathematical or analytical formula that describes the system's reliability. However, for a complex system, the determination of system reliability becomes a challenging task.

As the needs for complex systems in various industries employing modern technologies have grown rapidly during recent years, the amount of research work on the design for reliability has increased significantly. However, in most of the recent research conducted on the subject of design for reliability, field and test data were used as the main source to obtain component reliability. Moreover, only a part of a system (e.g., electrical or mechanical part) was studied, while hybrid electromechanical systems were not integrally analyzed. The following survey confirms these claims.

Dhillon and Rayapati (1986) presented a method to evaluate the reliability of complex systems composed of independent three-state devices. The method was demonstrated with the aid of a hypothetical example of a complex system. In addition, the impact of unit (i.e., three-state device) in the open mode and short mode failure probabilities on bridge network reliability was shown. Unlike simple systems, very few methodologies treat

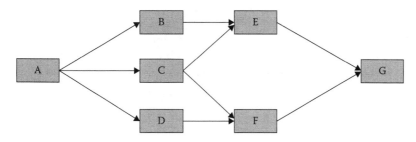

FIGURE 1.2
A complex system network.

the evaluation of the dependability of complex systems, especially those configured as networks, where it is difficult to take into consideration different links and factors that can affect the availability and reliability of such systems. Bayesian network (BN) is a very interesting tool facilitating the modeling of systems configured as network and the computation of marginal probabilities of the nodes of the system using prior and conditional probabilities. Torres-Toledano and Sucar (1998) proposed a general methodology to model the reliability of complex systems based on BN. In their work, a reliability structure represented as an RBD was first transformed to a BN representation. Then, the system reliability was obtained using some probability propagation techniques. This allows one to model complex systems that involve dependencies between the failures, which is difficult to obtain using conventional reliability analysis techniques. The relation between a BN and an FT along with some advantages of BN for modeling system reliability was also shown in their research.

Avontuur (2000) and Avontuur and van der Werff (2001) emphasized the importance of reliability analysis in the conceptual design phase, where it was demonstrated that it is possible to improve a design by applying reliability analysis techniques in the conceptual design phase. The aim was to quantify the costs of failure and unavailability and compare them with the investment cost to improve the reliability. Abo Al-Kheer et al. (2011) developed a model using the design for reliability approach to integrate the randomness of tillage forces into the design analysis of tillage machines, aiming at achieving reliable machines. The proposed approach was based on the uncertainty analysis of basic random variables and the failure probability of tillage machines. For this purpose, two reliability methods, namely, the MC simulation technique and the first-order reliability method, were utilized (O'Halloran et al., 2012).

Hryniewicz (2006) investigated the evaluation problem of system reliability using statistical data obtained from reliability tests of system components, in which the lifetimes of the components were described using an exponential distribution. He assumed that this lifetime data may be reported imprecisely and that this lack of precision may be described using fuzzy sets. As the direct application of the fuzzy sets methodology leads to very complicated and time-consuming calculations in this case, he proposed simple approximations of fuzzy numbers using shadowed sets introduced by Pedrycz (1998). Guo and Wilson (2013) proposed a Bayesian approach to assess the reliability of multicomponent systems. Their model allows evaluating system, subsystem, and component reliability using multilevel information. Data were collected over time and included binary, lifetime, and degradation data.

To reduce the cost of MC simulations for time-consuming processes, the Bayesian Monte Carlo (BMC) method was introduced by Rajabalinejad (2010). The BMC method reduces the number of realizations in MC according to the desired accuracy level. BMC also provides the possibility of considering more priors. In other words, different priors can be integrated into one model using BMC to further reduce the cost of simulations. He suggested speeding up the simulation process by considering the logical dependence of neighboring points as prior information. This information is used in the BMC method to produce a predictive tool through the simulation process.

In traditional methods proposed in reliability analysis, sometimes a complex system is considered as being composed of some subsystems in series, where the failure of any subsystem would be supposed to lead to the failure of the entire system. However, while the lifetimes of some subsystems are long enough and even never fail during the life cycle of the entire system, they are not influenced equally under different circumstances. In practice, such interferences will affect the model's accuracy, but it is seldom considered in traditional analysis. To address these shortcomings, Lin et al. (2011) presented a new approach for the reliability analysis of complex systems. They defined a certain fraction of the subsystems as a "cure fraction" under the consideration that the lifetimes of such subsystems are long enough during the life cycle of the entire system. By introducing environmental covariates and the joint power prior, their model was developed within the Bayesian survival analysis framework, and thus the problem for censored (or truncated) data in reliability tests was resolved.

Silvestri (2014) proposed a method to generate an exact analytical expression for the reliability of a complex system using a directed acyclic graph representing the system reliability block diagram. Additionally, he showed how statistical information stored in a reliability block diagram can be used to transform an analytical expression into a time-dependent function for system reliability.

Methods for computing the reliability of complex systems described by Utkin and Kozine (2001) were grounded on partial information on system components. They provided a tool for inferring the intervals as the natural extension and the upper and lower bounds of the characteristics to be interpreted as coherent upper and lower previsions. They utilized a generic algorithm to find a solution of the natural extension, breaking down the general problem into problems that are much easier to solve. In general, this can be made at the cost of a lesser precision in the previsions of interest. It was also shown that for some particular cases, the genuine and minimally coherent solutions can be found through the algorithm developed.

Reliability is an important phase in durable system designs, specifically in the early phase of product development. Soleimani et al. (2014) proposed a new methodology for complex system design for reliability. Specific test and field failure data scarcity was evaluated in their work as a challenge to implement design for the reliability of a new product. They employed the RBD method for modeling and simulation. The generic data were corrected to account for the design and environmental effects on the application.

The integral methodology evaluated the reliability of the system and assessed the importance of each component. In addition, the availability of the system was evaluated using MC simulation. As a case study, horizontal drilling equipment was used for the assessment of the developed method. In addition, Soleimani and Pourgol-Mohammad (2014) developed a methodology for the reliability evaluation of electromechanical systems. The method was applicable in the early design phase where there was only limited failure data available. When experimental failure data is scarce, generic failure data are searched from some related reliability data banks. In their method, RBD was used for modeling system reliability. An MC simulation technique was employed to simulate the system for reliability and availability calculations. The methodology contained the reliability importance analysis and reliability allocation to optimize the reliability.

Several methods such as decomposition, event space, and path tracing exist for analytically obtaining the reliability of a complex system (Ebeling, 1997). There are a number of advantages of using analytical techniques to determine system reliability, as opposed to the more common method of simulation. The primary advantage of the analytical solution is that a mathematical expression that describes the reliability of the system is obtained. Once the system's reliability function is determined, other analyses on the system can be performed. Such analyses include (1) determination of the system's probability density function, (2) determination of the warranty period, (3) determination of the system's failure rate, and (4) determination of the system's mean time to failure. In addition, optimization and reliability allocation techniques can be utilized to aid engineers in their design improvement efforts. Another advantage of using analytical techniques is the ability to perform static calculations and analyze systems with a mixture of static and time-dependent components. Finally, the reliability importance of components over time can be calculated with this methodology.

Complex systems are characterized by large numbers of components, cut sets or link sets, or by statistical dependence between the component states. These measures of complexity render the computation of system reliability a challenging task. Der Kiureghian and Song (2008) described a decomposition approach, which, together with a linear programming formulation, allowed determination of bounds on the reliability of complex systems with a manageable computational effort. The approach also facilitated multiscale modeling and analysis of a system, whereby varying degrees of detail could be considered in the decomposed system. The chapter also described a method for computing bounds on conditional probabilities by the use of linear programming, which can be used to update the system reliability for any given event. Applications to a power network demonstrated the methodology.

Full-system testing for large-scale systems is often infeasible or very costly. Thus, when estimating system reliability, it is desirable to employ a method that uses subsystem tests, which are often less expensive and more feasible.

Hill and Spall (2000) presented a method for bounding full-system reliabilities based on subsystem tests. The method did not require subsystems to be independent. It accounts for dependencies through the use of certain probability inequalities. The inequalities provide the basis for valid reliability calculations while not requiring independent subsystems or full-system tests. The inequalities allow for test information on pairwise subsystem failure modes to be incorporated, thereby improving the estimate of system reliability.

2

Fault Tree Analysis and Reliability Block Diagram

KEYWORDS: *complex system reliability, industrial robots, fault tree analysis (FTA), reliability block diagram (RBD).*

2.1 Introduction

As complex systems have become global and essential in today's society, the reliability of their design and the determination of their availability have become very important tasks for managers and engineers. Industrial robots are examples of these complex systems that are being increasingly used for intelligent transportation, production, and distribution of materials in warehouses and automated production lines. In this chapter, a comprehensive fault tree analysis (FTA) on the critical components of industrial robots is conducted. This analysis is integrated with the reliability block diagram (RBD) in order to investigate the robot system reliability. For practical implementation, a particular robot system is first modeled. Then, FTA is adopted to model the causes of failures, enabling the probability of success to be determined. In addition, RBD is employed to simplify the complex system of the robot for reliability evaluation purpose.

2.2 Problem Statement

The robot system under consideration consists of a drive unit, software control system, laser navigation system, safety system, attachments, batteries, brake system, steering system, and manual buttons. Among these subassemblies, the drive unit, usually a brushless DC electric motor, provides power for motion and operation. The laser navigation system, developed by Macleod and Chiarella (1993), is in essence a position measurement system to locate the robot. It comprises a rotating laser installed on the board of the robot and three beacons mounted along the border of the area to be covered. The safety system, with the aid of a laser detection system installed on the robot, is designed to avoid obstacles that could appear on the pathway. Attachments refer to those additional components that are used to assist in moving and carrying items, and batteries—usually the common lead–acid batteries—are used to supply power to the whole robot system.

2.3 Fault Tree Analysis

The robot is requested to distribute materials to multiple places in a system according to various requirements. Each time, once it receives an order, it optimizes the routes for completing the whole mission. In the proposed system of robots, satisfaction is reached when there is no failure during a mission. Therefore, investigating robot failures is necessary using the FTA. By inspecting the reason behind the undesired events that could happen in a system or during a mission, FTA allows one to trace back the root cause of a system or mission failure by using a systematic top-down approach. Moreover, the probability of system or mission failure can be computed via Boolean logic calculations with the aid of FTA. Attributed to that, FTA provides a straightforward and clear presentation to the reason behind various undesired events. Moreover, it supports both qualitative and quantitative analyses. FTA has been regarded as an effective, systematic, accurate, and predictive method to deal with safety and reliability problems in complex systems, such as safety issues in a nuclear power plant (Qureshi, 2007).

Here, the top evidence is the robot failure that encompasses failures of robot segments. In Figure 2.1, the fault tree of a robot is depicted. Also, the symbols employed in FTA are shown in Table 2.1.

In addition, to form an FTA and to analyze the causes of the failures, FTAs of all subsystems of robots are proposed. Therefore, as shown in Figure 2.1, "failure in allocation" needs to be focused on separately. Sending false commands, failure of control systems, and failure in the emergency system could be the technical reasons for such a fault. Loss of power supply is another important aspect that can cause malfunctioning of robots. Overheat, low electrical charge, leakage, and performance degeneration are some of the causes of failures. One of the major robot failures is failure during control and dispatching, which originates from the safety system.

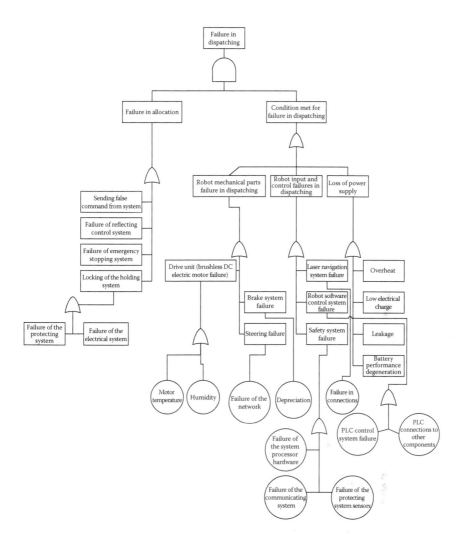

FIGURE 2.1
FTA of a robot.

2.4 Reliability Block Diagram

An RBD is a graphical representation of the components of a system and shows how they are reliability-wise related. The diagram represents the functioning state (i.e., success or failure) of the system in terms of the functioning states of its components. For example, a simple series configuration

TABLE 2.1

Symbols Being Used in the Proposed FTA

Symbols	Functions
Rectangle	Fault that is extracted from fault evidence via logic gates
Circle	Basic fault evidence
Triangle	The remainder of the tree is depicted elsewhere
OR gate	When any input evidence occurs, an output evidence takes place
AND gate	All input evidence should have an output evidence

indicates that all of the components must operate for the system to operate, and a simple parallel configuration indicates that at least one of the components must operate, and so on (Bistouni and Jahanshahi, 2014). An RBD performs the system reliability and availability analyses on large and complex systems using block diagrams to show network relationships. The structure of the RBD defines the logical interaction of failures within a system that are required to sustain system operation. The rational course of an RBD stems from an input node located at the left side of the diagram. The input node flows to arrangements of series or parallel blocks that conclude at the output node at the right-hand side of the diagram. A diagram should only contain one input and one output node. In the robot system under consideration, the symbols of failures shown in Table 2.2 are defined to configure the RBD.

Thereafter, the RBD corresponding to the proposed fault tree of the robot system is depicted in Figure 2.2.

A method is needed to compute the reliability of the system based on the faults and the RBD. In Chapter 3, the decision tree is proposed as the solution approach.

TABLE 2.2

Coding for Failures of a Robot System

B1	Failure in dispatching
B2	Failure in allocation
B3	Condition met for failure in dispatching
B4	Sending false command from system
B5	Failure of reflecting control system
B6	Failure of emergency stopping system
B7	Locking of the holding system
B8	Failure of the protecting system
B9	Failure of the electrical system
B10	Robot mechanical parts failure in dispatching
B11	Robot input and control failures in dispatching
B12	Loss of power supply
B13	Drive unit (brushless DC electric motor failure)
B14	Brake system failure
B15	Steering failure
B16	Laser navigation system failure
B17	Robot software control system failure
B18	Safety system failure
B19	Overheat
B20	Low electrical charge
B21	Leakage
B22	Battery performance degeneration
B23	Motor temperature
B24	Humidity
B25	Depreciation
B26	Failure of the network
B27	Failure in connections
B28	PLC connection to other components
B29	PLC control system failure
B30	Failure of the system processor hardware
B31	Failure of the communicating system
B32	Failure of the protecting system sensors

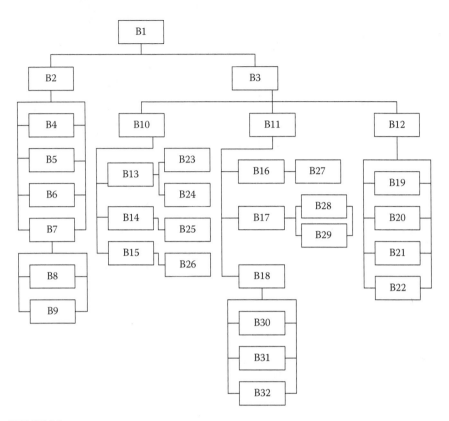

FIGURE 2.2
Reliability block diagram for a robot system.

3

Cost and Hazard Decision Tree Models

KEYWORDS: *industrial robot, cost decision tree, hazard decision tree.*

3.1 Introduction

In this chapter, the cost and hazard decision tree (C&HDT) is integrated for the first time, by adopting the fault tree analysis (FTA) and the reliability block diagram (RBD) from the previous chapter, in an approach to evaluate the reliability of a complex system called industrial robot. Some advantages of such a methodology include detailing of complex robot system components by FTA, providing a block diagram design for series/parallel presentation, using hazard function as a significant index in reliability and maintenance studies, and employing an efficient computational method of decision tree for overall evaluation. The proposed process can be easily implemented using the usual data collected by the maintenance department without any extra, time-consuming efforts. Here, the hazard decision tree (HDT) is configured to compute the hazard of each component and of the whole robot system. Through this research, a promising technical approach is established, allowing decision makers to identify the critical components of industrial robots along with their crucial hazard phases at the design stage.

3.2 Cost Decision Tree

A decision tree is used for the process of decision making process using a tree-like diagram of decisions and the corresponding possible consequences, including chance event outcomes, resource costs, and utility. A decision tree is a way of presenting an algorithm or algorithmic behavior. In this section, a cost-based decision tree is developed. Using the block diagram of a complex robot system proposed in Chapter 2 and introducing a couple of factors, namely, failure probability and repair cost, one can compute the minimal cost of maintaining system reliability.

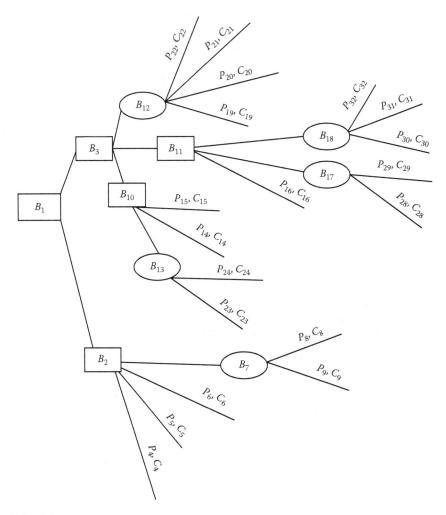

FIGURE 3.1
Cost decision tree.

Thus, assume P_i as the failure probability of component B_i and C_i as the repair cost for that component, then K_i is the total cost incurred to B_i, G_i is the minimal cost of repair for B_i, and Z is the minimal cost to maintain system reliability.

A configuration of the cost decision tree for a complex robot system is shown in Figure 3.1.

To compute mathematically,

$$K_{12} = P_{22} \cdot C_{22} + P_{21} \cdot C_{21} + P_{20} \cdot C_{20} + P_{19} \cdot C_{19}$$
$$K_{17} = P_{29} \cdot C_{29} + P_{28} \cdot C_{28}$$
$$G_{11} = \text{Min}\{K_{18}, K_{17}, K_{16}\}$$

$$K_{10} = P_{15} \cdot C_{15} + P_{14} \cdot C_{14} + K_{13}$$
$$K_{13} = P_{24} \cdot C_{24} + P_{23} \cdot C_{23}$$
$$G_3 = \text{Min}\{K_{12}, G_{11}, K_{10}\}$$

$$K_7 = P_8 \cdot C_8 + P_9 \cdot C_9$$
$$K_6 = P_6 \cdot C_6$$
$$K_5 = P_5 \cdot C_5$$
$$K_4 = P_4 \cdot C_4$$
$$G_2 = \text{Min}\{K_7, K_6, K_5, K_4\}$$
$$Z = \text{Min}\{G_3, G_2\}.$$

3.3 Hazard Decision Tree

The hazard function is a measure of the tendency to fail; the greater the value of the hazard function, the greater the probability of the impending failure. Technically, the hazard function is the probability of failure in a very small time interval, x_0 to $x_0 + dx$, given survival until x_0. It is also known as the instantaneous failure rate. Mathematically, the hazard function is defined as (Ebeling, 1997)

$$h(x) = \frac{f(x)}{R(x)}, \tag{3.1}$$

where
  $f(x)$ is the probability density function
  $R(x)$ is the reliability function
  $h(x)$ is the hazard function

Using expression 3.1 along with the following two expressions, if either the hazard function, reliability function, or probability density function is known, the remaining two functions can be derived:

$$R(x) = e^{-\int_{-\infty}^{x} h(\tau)d\tau}, \tag{3.2}$$

$$f(x) = h(x)e^{-\int_{-\infty}^{x} h(\tau)d\tau}. \tag{3.3}$$

Sometimes the hazard function (known as the failure rate) is defined as the ratio of the probability density function to the survival function $S(x)$ (Ebeling, 1997):

$$h(x) = \frac{f(x)}{S(x)} = \frac{f(x)}{1 - F(x)},$$

(3.4)

where $F(x)$ is the cumulative probability distribution function. The failure rate is defined for nonrepairable populations as the (instantaneous) rate of failure for the survivors to time t during the next instant of time. In the next instant, the failure rate may change and the units that have already failed play no further role since only the survivors count. Hence, the failure rate (or hazard rate) is denoted by $h(t)$ and is calculated as

$$h(t) = \frac{f(t)}{1 - F(t)} = \frac{f(t)}{R(t)} = \text{the instantaneous (conditional) failure rate.}$$

(3.5)

The failure rate is sometimes called "conditional failure rate" since the denominator $1 - F(t)$ (i.e., the population survivors) converts the expression into a conditional rate, given survival past time t. Since $h(t)$ is also equal to the negative of the derivative of $\ln R(t)$, we have the following useful identity (Kapur and Lamberson, 1977):

$$F(t) = 1 - e^{\left[-\int_0^t h(t)dt \right]}.$$

(3.6)

If we let

$$H(t) = \int_0^t h(t)dt$$

(3.7)

be the cumulative hazard function, then we have $F(t) = 1 - e^{H(t)}$. Accordingly, the other two useful identities that follow from these formulae are

$$h(t) = -\frac{d \ln R(t)}{dt},$$

(3.8)

$$H(t) = -\ln R(t).$$

(3.9)

If the hazard function increases, failures are caused by wear. If the hazard function decreases, infant mortality failures occur. Some causes of infant mortality failures include (1) improper use, (2) improper installation, (3) improper setup, (4) inadequate training, (5) poor quality control, (6) defective materials, (7) power surges, (8) inadequate testing, and (9) damage during storage or shipping (Mak, 2007).

It is also sometimes useful to define an average failure rate over any interval (T_1, T_2) that "averages" the failure rate over that interval. This rate, denoted by $AFR(T_1, T_2)$, is a single number that can be used as a specification or target for the population failure rate over that interval. The formula for calculating $AFR(T_1, T_2)$ values is (Ebeling, 1997)

$$AFR(T_2 - T_1) = \frac{\int_{T_1}^{T_2} h(t)\,dt}{T_2 - T_1} = \frac{H(T_2) - H(T_1)}{T_2 - T_1} = \frac{\ln R(T_1) - \ln R(T_2)}{T_2 - T_1}. \quad (3.10)$$

Moreover, if T_1 is 0, then it is dropped from the expression, that is,

$$AFR(0, T) = AFR(T) = \frac{H(T)}{T} = \frac{-\ln R(T)}{T}. \quad (3.11)$$

Based on these foundations, the hazard decision tree is configured as shown in Figure 3.2.

The notations used in Figure 3.2 are as follows:

$F_i(t)$: Cumulative probability distribution function for the failure of component B_i

$f_i(t)$: Probability density function for the failure of component B_i

$h_i(t)$: Hazard function of component B_i

$HT_i(t)$: Total hazard function of component B_i

X_i: The least hazard function value for component B_i

M: The least hazard function value for the robot

Using these notations, the following computation procedure is presented for the proposed hazard decision tree given in Figure 3.2:

$$h(t) = \frac{f(t)}{1 - F(t)}, \quad (3.12)$$

$$HT_{12}(t) = \frac{f_{22}(t)}{1 - F_{22}(t)} + \frac{f_{21}(t)}{1 - F_{21}(t)} + \frac{f_{20}(t)}{1 - F_{20}(t)} + \frac{f_{19}(t)}{1 - F_{19}(t)},$$

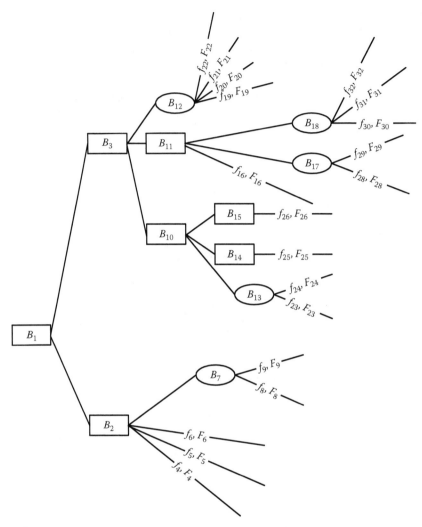

FIGURE 3.2
Hazard decision tree.

$$HT_{18}(t) = \frac{f_{32}(t)}{1 - F_{32}(t)} + \frac{f_{31}(t)}{1 - F_{31}(t)} + \frac{f_{30}(t)}{1 - F_{30}(t)},$$

$$HT_{17}(t) = \frac{f_{29}(t)}{1 - F_{29}(t)} + \frac{f_{28}(t)}{1 - F_{28}(t)},$$

$$HT_{16}(t) = \frac{f_{16}(t)}{1 - F_{16}(t)}.$$

Then, $X_{11} = \text{Min}\{HT_{18}(t), HT_{17}(t), HT_{16}(t)\}$. Moreover,

$$HT_{15}(t) = \frac{f_{15}(t)}{1 - F_{15}(t)},$$

$$HT_{13}(t) = \frac{f_{24}(t)}{1 - F_{24}(t)} + \frac{f_{23}(t)}{1 - F_{23}(t)},$$

$$HT_{14}(t) = \frac{f_{14}(t)}{1 - F_{14}(t)}.$$

Then,

$$X_{10} = \text{Min}\{HT_{13}(t), h_{15}(t), HT_{14}(t)\}$$
$$X_{3} = \text{Min}\{HT_{12}(t), X_{11}, X_{10}\}.$$

Furthermore,

$$HT_{17}(t) = \frac{f_{8}(t)}{1 - F_{8}(t)} + \frac{f_{9}(t)}{1 - F_{9}(t)},$$

$$h_{6}(t) = \frac{f_{6}(t)}{1 - F_{6}(t)},$$

$$h_{5}(t) = \frac{f_{5}(t)}{1 - F_{5}(t)},$$

$$h_{4}(t) = \frac{f_{4}(t)}{1 - F_{4}(t)}.$$

Consequently,

$$X_{2} = \text{Min}\{h_{6}(t), h_{5}(t), h_{4}(t), HT_{17}(t)\} \quad \text{and}$$
$$M = \text{Min}\{X_{3}, X_{2}\}.$$

An implementation study is conducted in the next section to demonstrate the application of the proposed methodology.

3.4 Implementation

The proposed hazard decision-making process can be implemented based on the depicted FTA and the corresponding RBD of the robot system (from Chapter 2). In the proposed case, the component failure rate and the probability distribution are presented in Table 3.1. Note that the failure rates are computed using data collected by the maintenance department of an industrial system. Moreover, the failure rates follow exponential and Weibull probability distributions according to the following probability distribution functions:

$$t \sim \text{EXP}(\theta) \Rightarrow f(t) = \theta e^{-\theta t}, \tag{3.13}$$

$$t \sim \text{WEB}(\alpha, \beta) \Rightarrow f(t; \alpha, \beta) = \frac{\alpha}{\beta^{\alpha}} t^{\alpha-1} e^{-\left(\frac{t}{\beta}\right)^{\alpha}}. \tag{3.14}$$

TABLE 3.1

Components and Their Failure Probability

Component	Probability Distribution	Hazard Value
B4	EXP(0.008)	0.008
B5	EXP(0.013)	0.013
B6	EXP(0.004)	0.004
B7	EXP(0.006)	0.006
B8	EXP(0.014)	0.014
B9	WEB(2.4, 520)	0.0056 for t = 600 (hours)
B14	EXP(0.004)	0.004
B15	EXP(0.003)	0.003
B16	EXP(0.005)	0.005
B19	EXP(0.002)	0.002
B20	EXP(0.004)	0.004
B21	EXP(0.004)	0.004
B22	WEB(4.7, 490)	0.036 for t = 700 (hours)
B23	EXP(0.006)	0.006
B24	EXP(0.003)	0.003
B25	WEB(1.7, 1600)	0.0011 for t = 1600 (hours)
B26	EXP(0.007)	0.007
B27	EXP(0.008)	0.008
B28	WEB(3.5, 385)	0.041 for t = 700 (hours)
B29	WEB(2, 950)	0.002 for t = 900 (hours)
B30	EXP(0.008)	0.008
B31	EXP(0.009)	0.009
B32	EXP(0.003)	0.003

Note also that the hazard function for the Weibull distribution depends on t and that, depending on whether α is greater than or less than 1, the hazard can increase or decrease with increasing t.

Using the hazard decision tree computation procedure given previously and using the data presented in Table 3.1, the following backward hazard computations are obtained:

$$H_{12}(t) = 0.036 + 0.004 + 0.004 + 0.002 = 0.046,$$
$$HT_{18}(t) = 0.003 + 0.009 + 0.008 = 0.02,$$
$$HT_{17}(t) = 0.041 + 0.002 = 0.043,$$
$$X_{11} = \text{Min}\{0.02, 0.043, 0.005\} = 0.005;$$
$$HT_{13}(t) = 0.003 + 0.006 = 0.009,$$
$$X_{10} = \text{Min}\{0.009, 0.004, 0.003\} = 0.003;$$
$$X_3 = \text{Min}\{0.046, 0.015, 0.003\} = 0.003;$$
$$HT_{17}(t) = 0.014 + 0.0056 = 0.0196,$$
$$X_2 = \text{Min}\{0.004, 0.013, 0.008, 0.0196\} = 0.004;$$
$$M = \text{Min}\{0.004, 0.003\} = 0.003.$$

Therefore, the hazard (failure) rate of the robot system according to its component failure probabilities is 0.3%. Thus, the maintenance department would try to reduce this hazard rate to increase the system's reliability.

3.5 Discussions

In this chapter, due to multiple components of the system, a decision tree–based hazard function was proposed to evaluate the overall failure rate of the system integrated with the FTA–RBD model. Using a procedure for backward computation in a hazard decision tree, the total hazard rate of the system was obtained. The advantages of such a methodology include detailing of complex robot system components by FTA, providing a block diagram design for series/parallel presentation, using hazard function as a significant index in reliability and maintenance studies, and employing an efficient computational method of decision tree for overall evaluation. As future research directions, developing replacement models for robot components, multistate system design, and repair cost models is suggested.

4

Binary State and Bernoulli Trials Reliability Model

KEYWORDS: *reliability assessment, autonomous robot, binary state system, Bernoulli trials.*

4.1 Introduction

Availability of a system is a crucial factor for planning and optimization. The concept is more challenging for modern systems such as robots and autonomous systems consisting of a complex configuration of components. In this chapter, a reliability evaluation framework is developed for a system of binary state autonomous robots in an automated manufacturing environment. In this framework, the concept of binary state reliability model is developed. Due to inefficacy of the method for a larger number of components involved in complex systems, an extension of the Bernoulli trials approach is proposed. In an implementation study, the effectiveness and computational efficiency of the proposed method are illustrated.

4.2 Problem Statement

Consider a large-scale manufacturing system including automated processes and multiple autonomous robots. In this system, appropriate functioning of the facilities is guaranteed by the functioning of their vital equipment. The process of evaluation and analysis of the system performance based on the system availability is significant. In this regard, reliability evaluation comes into the picture as an effective instrument. While the system under consideration is complex due to having many material handling autonomous robots and manufacturing machines, developing an efficient approach to compute and analyze system reliability has a significant benefit. For instance, this helps in the proper arrangement of autonomous robots and machine layout (robots and machines are considered components) as well as in specifying an adequate process plan. Note that all these factors influence the reliability of the whole system. In addition, as the robots are assumed to operate in two states, working and not working, some techniques that are capable of determining the reliability of a binary state complex system are explained in the next section.

4.3 Binary Model

Generally, complex systems are systems having more than 25 components, and thus, in some cases, constructing a correct conditional probability model may be difficult for the purpose of reliability computation. Here, some techniques that are fully capable of providing exact reliabilities of such systems as well as determining the reliabilities of smaller systems are described. The process of evaluating the reliabilities of large systems requires algorithms that are based on binary decision systems, which, during the last decade, have been established as state-of-the-art techniques.

Reliability is defined as the probability that an element (i.e., a component, a subsystem, or a full system) will accomplish its assigned task within a specified time designated as the interval, $t \in [0, t_M]$. Assume operational and failed elements are in the 1 and 0 states, respectively. Furthermore, only coherent systems are considered in this chapter. A coherent system is one in which (a) the reliability of the system increases if the reliability of its components increases and (b) there are no irrelevant components. The failure of any component or a set of components in a coherent system cannot cause an increase in reliability, and every component has some effect, however small, on the overall reliability (Ebeling, 1997). Let p_i denote the reliability of the ith component of a system. Then, this component has an unreliability q_i defined in Equation 4.1:

$$q_i = 1 - p_i. \tag{4.1}$$

In other words, as the component is always in either one of the two possible states (operational or failed), then

$$p_i + q_i = 1. \tag{4.2}$$

To perform quantitative system reliability analysis, it is necessary to ascribe a probability that the individual components are either operational or have failed. In this case, a reliability function, also called a survival function, defines the probability that the component will perform its intended task (usually subject to some stated set of environmental conditions, such as vibration and temperature) for some specified performance period. The performance period may be a function of cycles, distance, or time. Although the techniques presented here can employ a reliability function that depends on any of these three parameters, that is, cycles, distance, and time, the focus is on determining the probability of system failure as a function of time. It is also critical that the reliabilities are estimated in a legitimate and appropriately conservative fashion. Several different functions have been used in the literature to characterize the probability distribution of failures as a function of time. Some of the more common reliability functions include the exponential, normal, log-normal, and Weibull distributions. In this work, however, the exponential probability distribution is used. The exponential distribution is appropriate for components with a failure rate that is time independent. Most electronic devices and also autonomous robots demonstrate such a constant failure rate during their useful lifetime, which is the time following a "burn in" that eliminates any weak or faulty components (Ebeling, 1997). The reliability function for a single-component system associated with the exponential distribution is

$$r\left(\lambda, t\right) = e^{-\lambda t},\qquad(4.3)$$

where $r(\lambda, t)$ is the probability that a component with a failure rate λ will be operational at time t.

4.4 Extension of the Bernoulli Trials Approach

In this section, an extension of the Bernoulli trials approach is proposed for the binary state complex system reliability computations. Consider a random experiment with two possible outcomes with probabilities p and q, where $p+q=1$. Now consider a compound experiment involving a sequence of n independent repetitions of this experiment. Such a sequence is known as a sequence of Bernoulli trials. Let 0 denote failure and 1 denote success in each experiment. Moreover, let S_n be the sample space of the experiment involving n Bernoulli trials, defined by $S_1 = \{0, 1\}$, $S_2 = \{(0,0), (0,1), (1,0), (1,1)\}$, ... , $S_n = \{2^n n - \text{tuples of 0's and 1's}\}$. While the probability assignment over the sample space S_1 is already specified as $P(0) = q$, $P(1) = p$, $p+q=1$, we wish to assign probabilities to the points in S_n. To this aim, let A_i and A_i' be the success and failure on trial i, respectively. Then, $P(A_i) = p$ and $P(A_i') = q$. Consider s as an element of S_n such that $s = (1, 1, \ldots, 1,\ 0, 0, \ldots, 0)$, [$k$ 1's and $(n-k)$ 0's]. Therefore,

$$P(s) = p^k q^{n-k}.\qquad(4.4)$$

Similarly, any sample point with k 1's and $(n - k)$ 0's is assigned the probability $p^k q^{n-k}$. Noting that there are $\binom{n}{k}$ such points, the probability of achieving exactly k successes in n trials is obtained using the binomial distribution

$$P(S_n) = \binom{n}{k} p^k q^{n-k}, \quad k = 0, 1, \ldots, n. \tag{4.5}$$

Now, consider a system with n components that requires at least $m (\leq n)$ components to function for the correct operation of the system (the m-out-of-n system). Note that if we let $m = n$, then we have a series system, and if we let $m = 1$, then we have a system with parallel redundancy. Assume the n components are statistically identical and function independently of each other. Let R denote the reliability of a component (and $Q = 1 - R$ gives its unreliability). Then, the experiment of observing the status of n components can be thought of as a sequence of n Bernoulli trials, each with the probability of success equal to R. Now, the reliability of such a system denoted by $R_{m|n}$ is the probability of exactly i components functioning properly, that is, $(i = m, m+1, \ldots, n)$. It is obtained by

$$R_{m|n} = \sum_{i=m}^{n} p(i) = \sum_{i=m}^{n} \binom{n}{i} R^i (1-R)^{n-i}. \tag{4.6}$$

Thus, Equation 4.6 is appropriate for computing the reliability of the proposed complex system of autonomous robots. Any calculating device or software can compute the system reliability even for large m.

4.5 Implementation

Robots transfer products from machine to machine through guide-paths. Due to several tasks that autonomous robots have to fulfill and the deadlock that occurs when two or more robots choose the same path to move, the reliability analysis of the system is significant. To facilitate the method introduced and to simplify the calculations, a couple of assumptions are made for reliability modeling: (1) components have the same lifetime distribution and (2) components that fail rarely do not lead to system failure.

In a numerical study, 25 components are considered to be either active (1) or inactive (0), and the failure rate follows an exponential distribution at $\lambda = 0.03$. In other words, the reliability of each component is computed using Equation 4.3. Here, we consider a case that four of the robots and machines encounter failure for $t = 25$ hours. Hence, using Equation 4.6, we obtain 0.000165 as the reliability of the whole system when at least 21 components are active and working properly.

4.6 Discussions

In this chapter, a novel approach was presented to evaluate the reliability of a complex system, including autonomous robots, in a manufacturing environment. The proposed model considered binary state robots that function as material handling devices in a manufacturing system. In this approach, the reliability functional block diagram was presented, based on which the truth table shows the probability of failure for the components. Then, the sum of states method was employed to compute the reliability of a complex system with a smaller number of components. With respect to polynomial equations derived to compute system reliability, an extension of the Bernoulli trials approach was suggested for a complex system with a larger number of components. At the end, the censored data were analyzed to estimate the failure rate for effective planning by the maintenance department for a higher level of reliability.

As for future research directions, the following can be pointed out:

- Considering multiple state system of a robot instead of using a binary state
- Considering state transition in the modeling and evaluation of reliability for the proposed robotic system
- Developing a predictive maintenance plan in parallel with reliability evaluation

5

Binary Decision Diagram
Reliability Model

KEYWORDS: *binary decision diagram (BDD), automation, complex system, reliability model.*

5.1 Introduction

Assume a production system having multiple robots and work stations leading to a complex system. To evaluate the reliability of this system, a network of components is considered configuring a complex reliability network. A robotic network is considered, which has perfect vertices and imperfect links. It means path links may fail with known probability. We obtain the reliability of the given network by using an exact method and with a binary decision diagram (BDD). BDD-based reliability evaluation involves three main steps. First, order the given path link. Second, generate the reliability function with the help of mini paths from source to sink. Finally, apply the Shannon decomposition formula to compute the reliability of the given network.

5.2 Problem Specification

The reliability evaluation under study is in two states—working and breakdown. Then, multiple robots are considered in the two states. To handle such a complex problem in a network-structured production system, a BDD is proposed, the reasons for which are as follows: adaptability with the network system, Boolean decision systems according to the two states of the robots, flexible computational capability with increase of size of the system components. In the proposed BDD, the working state is shown by 1 and breakdown (stop working) is shown by 0. More details are given in the next section.

An efficient method for generating the BDD of a complex robotic system is proposed. The reliability is evaluated via the BDD and by applying a truth table and the Shannon decomposition formula. Considering the binary decision tree, the computations of reliability are presented.

5.3 Modeling

Akers (1978) first introduced BDD to represent Boolean functions, that is, a BDD is a data structure used to represent a Boolean function. Bryant (1992) popularized the use of BDD by introducing a set of algorithms for efficient construction and manipulation of the BDD structure. The BDD structure provides compact representations of Boolean expressions. A BDD is a directed acyclic graph (DAG) based on the Shannon decomposition. The Shannon decomposition for a Boolean function is defined as follows:

$$f = l \cdot f_{l=1} + l \cdot f_{l=0} \qquad (5.1)$$

where
  l is one of the decision variables
  f is the boolean function evaluated at $l = i$

By using the Shannon decomposition, any Boolean expression can be transformed into binary DAG. BDD is used to work out the terminal reliability of the links. The maintenance department records the failures, and thus probability distribution is obtained leading to reliability evaluation. Since we have a network of components in the production system that is active or inactive (binary state), the binary decision tree is configured. The following notations are used for this purpose:

  L_j: robot jth
  F_u: robot performance function
  K: counter for robots; $j = 1, 2, ..., K$

5.4 Implementation

Let us consider an example for the implementation of the proposed robotic system reliability evaluation. A production system having nine work stations and four robots to process material handling in considered. The movement paths of the robots are in series and parallel with respect to the sequence of work

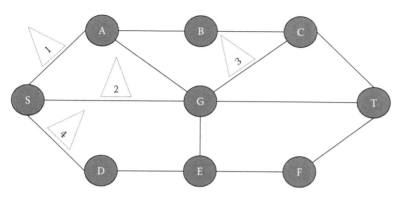

FIGURE 5.1
Schematic representation of the robotic system.

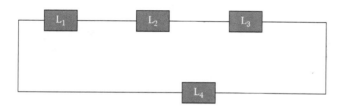

FIGURE 5.2
Reliability block diagram.

stations they service. If two robots have a common path, except the source (S) and sink (T) nodes, then they are in series; otherwise, they are in parallel. A schematic representation of the system network is shown in Figure 5.1.

According to the series/parallel structure of the components, the corresponding reliability block diagram is drawn as shown in Figure 5.2.

i		Robot ith
L_1, L_4	⟶	Parallel
L_2, L_4	⟶	Parallel
L_1, L_2	⟶	Series
L_1, L_3	⟶	Series
L_2, L_3	⟶	Series
L_3, L_4	⟶	Parallel

Also, the corresponding truth table for the four-robot system is given in Table 5.1.

TABLE 5.1

Truth Table for the Four-Robot System

L_1	L_2	L_3	L_4	f_u
0	0	0	0	0
1	0	0	0	0
0	1	0	0	0
0	0	1	0	0
0	0	0	1	1
1	1	0	0	0
1	0	1	0	0
1	0	0	1	1
0	1	1	0	0
0	1	0	1	1
0	1	1	1	1
0	0	1	1	1
1	1	1	0	1
1	0	1	1	1
1	1	0	1	1
1	1	1	1	1

According to the truth table, a binary decision tree is formed and shown in Figure 5.3.

While there are two states, working (S) and failure (F), for each robot, 16 cases are formed as given here. Since robots are similar, the failure rate is considered 0.02 for all of them, that is, $p(L1) = p(L2) = p(L3) = p(L4) = 0.02$. Finally, the reliability of each case considering the binary decision tree and Shannon decomposition formula is as follows:

$$L1_F\, L2_F\, L3_F\, L4_F\colon 1-\left[\left(1-\left(0.02*0.02*0.02\right)\right)*\left(1-0.02\right)\right]=0.020007$$

$$L1_S\, L2_F\, L3_F\, L4_F\colon 1-\left[\left(1-\left(0.98*0.02*0.02\right)\right)*\left(1-0.02\right)\right]=0.02038416$$

$$L1_F\, L2_S\, L3_F\, L4_F\colon 1-\left[\left(1-\left(0.02*0.98*0.02\right)\right)*\left(1-0.02\right)\right]=0.02038416$$

$$L1_F\, L2_F\, L3_S\, L4_F\colon 1-\left[\left(1-\left(0.02*0.02*0.98\right)\right)*\left(1-0.02\right)\right]=0.02038416$$

$$L1_F\, L2_F\, L3_F\, L4_S\colon 1-\left[\left(1-\left(0.02*0.02*0.02\right)\right)*\left(1-0.98\right)\right]=0.98000016$$

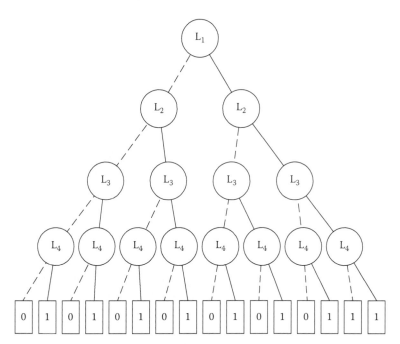

FIGURE 5.3
Binary decision tree for the four-robot production system.

$$L1_S\ L2_S\ L3_F\ L4_F{:}\ 1-\Big[\big(1-(0.98*0.98*0.02)\big)*(1-0.02)\Big]=0.03882384$$

$$L1_S\ L2_F\ L3_S\ L4_F{:}\ 1-\Big[\big(1-(0.98*0.02*0.98)\big)*(1-0.02)\Big]=0.03882384$$

$$L1_S\ L2_F\ L3_F\ L4_S{:}\ 1-\Big[\big(1-(0.98*0.02*0.02)\big)*(1-0.98)\Big]=0.98000784$$

$$L1_F\ L2_S\ L3_S\ L4_F{:}\ 1-\Big[\big(1-(0.02*0.98*0.98)\big)*(1-0.02)\Big]=0.03882384$$

$$L1_F\ L2_S\ L3_F\ L4_F{:}\ 1-\Big[\big(1-(0.02*0.98*0.02)\big)*(1-0.02)\Big]=0.02038416$$

$$L1_F\ L2_S\ L3_S\ L4_S{:}\ 1-\Big[\big(1-(0.02*0.98*0.98)\big)*(1-0.98)\Big]=0.98038416$$

$$L1_F\ L2_F\ L3_S\ L4_S{:}\ 1-\Big[\big(1-(0.02*0.02*0.98)\big)*(1-0.98)\Big]=0.98000784$$

$$L1_S\ L2_S\ L3_S\ L4_F{:}\ 1-\Big[\big(1-(0.98*0.98*0.98)\big)*(1-0.02)\Big]=0.94236816$$

$$L1_S \, L2_F \, L3_S \, L4_S: 1 - \left[\left(1 - (0.98 * 0.02 * 0.98)\right) * (1 - 0.98) \right] = 0.98038416$$

$$L1_S \, L2_S \, L3_F \, L4_S: 1 - \left[\left(1 - (0.98 * 0.98 * 0.02)\right) * (1 - 0.98) \right] = 0.98038416$$

$$L1_S \, L2_S \, L3_S \, L4_S: 1 - \left[\left(1 - (0.098 * 0.98 * 0.98)\right) * (1 - 0.98) \right] = 0.99882384$$

This way, the reliability of a complex robotic system is computed.

5.5 Discussions

Reliability assessment in the current advanced manufacturing systems is significant due to the large amount of economic investments. Also, a complex system (network) of components motivates developing efficient methods for performance analysis. Binary state systems are the simplest, so they are extensively considered in production systems due to maintenance consideration. Combining stochastic process computations such as branching processes with high-performance artificial intelligence techniques helps the decision makers to obtain more reliable results. Research in this field is toward the application of neural network methodologies for computing reliability in stochastic and uncertain components, specifically in automation applications.

6

Decomposition and Minimal Path and Cuts Method

KEYWORDS: *complex system reliability, industrial robots, decomposition method (DM), minimal path and cuts method (MPCM).*

6.1 Introduction

As complex systems have become global and essential in today's society, their reliable design and the determination of their availability have become very important aspects for managers and engineers. Industrial robots are examples of these complex systems that are being increasingly used for intelligent transportation, production, and distribution of materials in warehouses and automated production lines. In this chapter, two techniques of reliability evaluation are developed for a complex system of robots. The decomposition method (DM) and the minimal path and cuts method (MPCM) are adapted for the proposed complex system. For practical implementation, a particular robot system is first modeled. Then, a reliability block diagram is adapted to model the complex system for the purpose of reliability evaluation.

6.2 Proposed Problem

Consider a complex production system including automated processes and multiple robots. In this system, appropriate functioning of the facilities is guaranteed by the functioning of their vital equipment. The process of evaluation and analysis of the system performance based on the system availability is significant. In this regard, reliability evaluation comes into the picture as an effective instrument. While the system under consideration is complex due to having many material handling robots, developing an efficient approach to compute and analyze system reliability has a significant benefit. For instance, this helps in the proper arrangement of robots and machine layout as well as in specifying an adequate process plan. Note that all these factors influence the reliability of the whole system. In addition, as the robots are assumed to operate in two states, working and not working, some techniques that are capable of determining the reliability of a complex system are explained in the next section (Shojaeifar et al., 2016).

6.3 Decomposition Method

One way to determine the reliability of a complex system is the DM. According to the DM, a component is chosen close to the left or to the right end of the system block diagram. This component is called "keystone." Then, the conditional system reliabilities, when the "keystone" survives and fails, are computed, respectively. The reliability of the whole system is then determined as a weighted average of these two conditional reliabilities, where the weights are the reliability of the "keystone," R and $1 - R$, respectively (Shojaeifar et al., 2016).

Let us denote the "keystone" by C_X and the corresponding reliability by R_X:

$$R_{SYS} = R_X \cdot R_{SYS|C_X} + (1 - R) \cdot R_{SYS|\bar{C}_X}, \tag{6.1}$$

where

$R_{SYS|C_X}$ is the conditional reliability when the "keystone" survives
$R_{SYS|\bar{C}_X}$ is the conditional reliability when the "keystone" fails

6.4 Minimal Paths and Cuts Method

The MPCM is a technique for reliability computation in complex systems, specifically network systems. Let $\psi(x_1, \ldots, x_n)$ be a function of n variables, $0 \le x_i \le 1$ for all $i = 1, \ldots, n$. This function is called a "structure function" if $\psi(I_1, \ldots, I_n) = 1$ when the system survives and is equal to zero otherwise, where $I_1, \ldots, I_n$ are survival indicators of the components (Shojaeifar et al., 2016).

Consider a system S of n components, represented by a given structure function. Let $P = \{C_{i1}, \ldots, C_{im}\}$ be a set of m components of S. P is called a "path set" if the system S survives when all the elements of P survive. A path set, P, is called minimal if it is not a path set following the exclusion of any of its members, that is, no proper subset of P is a path set. When the block diagram of a system is obtained, all the minimal paths can be listed.

A "cut set" is a set of components of a system such that if all the components belonging to the set fail, then the system also fails. A cut set is called minimal if the survival of any of its elements entails system survival (Shojaeifar et al., 2016).

6.5 Implementation

Let us consider a numerical example to show the effectiveness of implementing the proposed methods for the reliability evaluation of an advanced production system. Consider a production system having nine robots for material handling, which are either active or failed (binary state). The robots are unidirectional and move between stations S and T. A configuration of the system is drawn in Figure 6.1.

Using the DM, we need to determine a keystone robot first. Robot 6 is considered as keystone, and using Equation 6.2, the reliability of the system is computed:

$$R_{sys} = R_6 \cdot R_{sys}|R_6 + (1 - R_6) R_{sys}|R_{\bar{6}}. \tag{6.2}$$

Since the formula is conditional, we need to compute the reliability separately based on the given condition. First, consider that if robot 6 is active, then all paths including robot 6 are functioning. Thus, the paths are

$$1\text{-}2\text{-}6\text{-}8 \quad 3\text{-}2\text{-}6\text{-}8 \quad 1\text{-}2\text{-}6\text{-}7\text{-}9 \quad 3\text{-}2\text{-}6\text{-}7\text{-}9$$

The series/parallel structures in different paths are taken into account for the reliability evaluation of the system. Robots 1 and 2 are in series and their integration is in parallel with robot 3. Also, robots 7 and 9 are in series and their integration is in parallel with robot 8, and the integration is in series with robot 6. Then, mathematically we have

$$R_6 \cdot R_{sys}|R_6 = 1 - \left[1 - (R_1 \cdot R_2) \cdot (1 - R_3)\right] \cdot R_6 \cdot \left[\left(1 - (1 - R_8)\right)\left(1 - (R_7 \cdot R_9)\right)\right]. \tag{6.3}$$

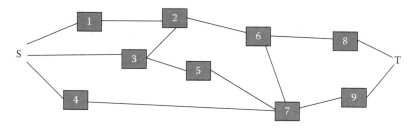

FIGURE 6.1
Configuration of the system under study.

The second case is when the system is conditionally not working, so robot 6 as a keystone is failed. Then, the paths in which robot 6 is not present should be selected. The paths are

$$3\text{-}5\text{-}7\text{-}9 \quad 4\text{-}7\text{-}9$$

Then, according to the DM formula, we obtain

$$(1 - R_6) \cdot R_{sys} | R_{\bar{6}} = \left[1 - \left(\left(1 - (R_3 \cdot R_5) \right) \left(1 - (R_4 \cdot R_7) \right) \right) \right] \cdot R_9 \cdot (1 - R_6). \quad (6.4)$$

The reliability of the system is the sum of the two preceding relations as follows:

$$R_{sys} = 1 - \left[1 - (R_1 \cdot R_2) \cdot (1 - R_3) \right] \cdot R_6 \cdot \left[\left(1 - (1 - R_8) \right) \left(1 - (R_7 \cdot R_9) \right) \right]$$
$$+ \left[1 - \left(\left(1 - (R_3 \cdot R_5) \right) \left(1 - (R_4 \cdot R_7) \right) \right) \right) \cdot R_9 \cdot (1 - R_6) \right]. \quad (6.5)$$

Since robots are similar and homogeneous ($R = 0.845$), the numerical result of system reliability that is obtained when robot 6 is the keystone is 0.928.

The DM is used when the system is in network configuration.

Now, consider a system having the reliability block diagram shown in Figure 6.2.

Also, the block diagram coding is presented in Table 6.1.

According to the block diagram of a nine-robot system, the reliability of the whole system is computed using MPCM. Initially, the existing minimal paths are determined. Note that L_j implies robot jth.

Path 1: $\{L_1, L_2, L_6, L_8\}$
Path 2: $\{L_3, L_5, L_7, L_9\}$
Path 3: $\{L_4, L_7, L_9\}$
Path 4: $\{L_3, L_2, L_6, L_8\}$
Path 5: $\{L_3, L_2, L_6, L_7, L_9\}$
Path 6: $\{L_1, L_2, L_6, L_7, L_9\}$

In identifying the minimal paths, it should be noted that robot movement is unidirectional. Then the proposed structure function is

$$\Phi(I_1, \ldots, I_9) = \Phi\left(I_1 I_2 I_6 I_8, I_3 I_5 I_7 I_9, I_4 I_7 I_9, I_3 I_2 I_6 I_8, I_3 I_2 I_6 I_7 I_9, I_1 I_2 I_6 I_7 I_9 \right)$$
$$= I_1 I_2 I_6 I_8 \cdot I_3 I_5 I_7 I_9 \cdot I_4 I_7 I_9 \cdot I_3 I_2 I_6 I_8 \cdot I_3 I_2 I_6 I_7 I_9 \cdot I_1 I_2 I_6 I_7 I_9 \quad (6.6)$$
$$= I_1 I_2 I_3 I_4 I_5 I_6 I_7 I_8 I_9.$$

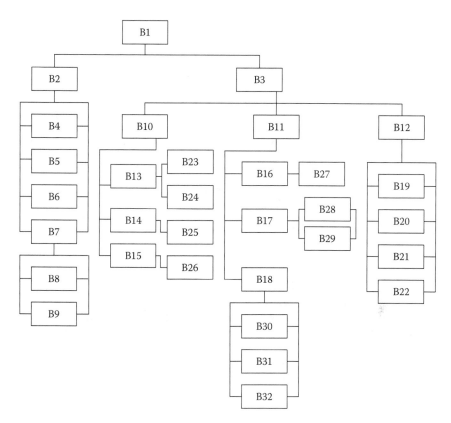

FIGURE 6.2
Reliability block diagram.

While robots are only in two states, activity and failure, then *I*s just take the value of 0 or 1. So, the reliability of the system is

$$R_{SYS} = R_1R_2R_6R_8 \cdot R_3R_5R_7R_9 \cdot R_4R_7R_9 \cdot R_3R_2R_6R_8 \cdot R_3R_2R_6R_7R_9 \cdot R_1R_2R_6R_7R_9$$
$$= R_1R_2R_3R_4R_5R_6R_7R_8R_9. \tag{6.7}$$

Since we consider homogeneous robots having similar reliability, the reliability of the system is

$$R_{SYS} = (0.845)^9 = 0.22.$$

TABLE 6.1

Block Diagram Coding

B1	Failure in dispatching
B2	Failure in allocation
B3	Condition met for failure in dispatching
B4	Sending false command from system
B5	Failure of reflecting control system
B6	Failure of emergency stopping system
B7	Locking of the holding system
B8	Failure of the protecting system
B9	Failure of the electrical system
B10	Robot mechanical parts failure in dispatching
B11	Robot input and control failures in dispatching
B12	Loss of power supply
B13	Drive unit (brushless DC electric motor failure)
B14	Brake system failure
B15	Steering failure
B16	Laser navigation system failure
B17	Robot software control system failure
B18	Safety system failure
B19	Overheat
B20	Electrical charge
B21	Leakage
B22	Battery performance degeneration
B23	Motor temperature
B24	Humidity
B25	Depreciation
B26	Failure of the network
B27	Failure in connections
B28	PLC connection to other components
B29	PLC control system failure
B30	Failure of the system processor hardware
B31	Failure of the communicating system
B32	Failure of the protecting system sensors

Now, the cut paths are determined using the block diagram. Cut paths are the ones that fail when all the components on them fail. The cut paths are listed as follows:

$$\{L_1, L_3, L_6\}; \quad \{L_1, L_3, L_7\}; \quad \{L_2, L_3, L_4\}; \quad \{L_2, L_3, L_7\}; \quad \{L_2, L_5, L_4\};$$

$$\{L_2, L_5, L_7\}; \quad \{L_6, L_3, L_4\}; \quad \{L_6, L_3, L_7\};$$

$$\{L_6, L_5, L_4\}; \quad \{L_6, L_5, L_7\}; \quad \{L_6, L_9\}; \quad \{L_8, L_9\}; \quad \{L_2, L_9\}; \quad \{L_7, L_8\}.$$

As a result, the numerical value of reliability is obtained to be 0.8741.

6.6 Discussions

In this chapter, two methods were adapted to evaluate the reliability of a complex production system including robots. The proposed model considered binary state robots that function as material handling devices in production systems. In this approach, the reliability block diagram was presented, based on which the structure of robots was determined. Then, using the DM and the MPCM, the reliability of the complex system was computed. As for future research directions, the following can be pointed out:

- Considering multiple state system of a robot instead of using a binary state
- Considering state transition in the modeling and evaluation of reliability for the proposed robotic system
- Developing a predictive maintenance plan in parallel with reliability evaluation

7

Bayesian Network Reliability Model

KEYWORDS: *Bayesian network (BN), complex system, reliability evaluation, industrial robots.*

7.1 Introduction

In many researches, the evaluation of reliability and availability is generally focused on simple cases of series and parallel systems or components. But, in real industrial environments, specifically automated systems, the components are configured in a network structure, where the interactions between the components are defined by dependability links making a complex system. Thus, an appropriate method to handle such a problem is the Bayesian network (BN), which is a useful tool to present both qualitative and quantitative representations of the links between the components. The structure of the network reflects the conditional dependencies among the links and components using prior and posterior probabilities.

7.2 Bayesian Network

A BN is a probabilistic graphical model that represents a set of random variables and their conditional dependencies via a directed acyclic graph. In BNs, nodes show random variables that represent robot failures in our case. Links imply the conditional dependencies between any two nodes. Each node is associated with a probability function that takes a particular set of values for the node's parent variables and gives the probability of the variable represented by the node.

7.3 Bayesian Reliability Model

In a production system having several robots to function, the failure of each robot influences the others. Then, using the probability of failure and employing BN, one can compute the reliability of the whole system. To model the proposed system as a BN, the nodes and links and the corresponding prior probabilities are first determined. For the prior probabilities, data collected by the maintenance department and experts' opinions are helpful.

If $P(A)$ is the prior probability and $P(B)$ is the occurrence probability, then $P(A|B)$ is the posterior probability. Thus, the Bayes relation is given as follows:

$$P(A|B) = \frac{P(B|A) \cdot P(A)}{P(B)}. \tag{7.1}$$

Using the block diagram, the Bayesian influence diagram can be drawn to find the dependencies of the failure of components, in order to compute the reliability of the system.

The Bayesian theory is essentially based on the mathematical concept of probability. For a BN, we have to establish a probability table for each node— the prior probability tables for the root nodes and the conditional probability tables for the other nodes. For the prior probability tables, we can use the following relation:

$$A = \frac{\text{Available time}}{\text{Available time} + \text{Unavailable time}}. \tag{7.2}$$

So, the probability of a given root node "x" is given in Table 7.1.

The values 0 and 1 in the table imply the binary state system, 1 for the operation state and 0 for the failure state. For the other BN nodes, the conditional probability concept is applied. So, the availability of a given node has to be evaluated by knowing the state of its parent nodes.

Then, we can configure conditional probabilities for node x in Table 7.2, when we are aware of the condition of node y. In fact, nodes x and y fail (0) or work (1) and they are available or unavailable, respectively.

TABLE 7.1

Prior Probability Table of a Root Node "x"

x	0	1
$P(x)$	$1 - A$	A

TABLE 7.2

Conditional Probability of a Node "x" Knowing the State of Its Parent Node "y"

	x	
y	0	1
0	$1 - P_{X|Y} \cdot A$	$P_{X|Y} \cdot A$
1	$1 - A$	A

Using historical data and conditional probability computations, posterior probabilities are obtained, and thus, applying backward calculations of BN and the law of total probability, the reliability of the system is computed.

7.4 Implementation

To show the applicability of the proposed model, an implementation study is presented. A nine-robot production system is considered. A configuration of the system is shown in Figure 7.1. The conditional reliability evaluation is done due to the occurrence of deadlock caused by a failed robot moving in the paths of other robots.

To emphasize the relation and dependency of the robots, the Bayesian influence diagram is depicted as shown in Figure 7.2.

In the BN of Figure 7.2, robots are numbered from 1 to 9, S is the initial station, and T is the final station for a robot to visit. For instance, if robot 1 fails and is unavailable, then robot 2 confronts a deadlock and the probability is $P(1|2)$. For availability (A) calculation, consider an eight-hour daily shift multiplied by 300 working days, in which there is one-hour unavailability for deadlock resolution in each daily shift:

$$A = \frac{7 \times 300}{(1 \times 300) + (7 \times 300)} = 0.875.$$

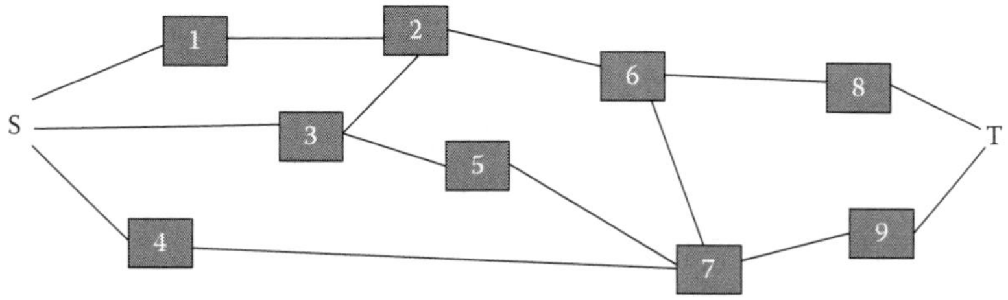

FIGURE 7.1
Block diagram of a nine-robot production system.

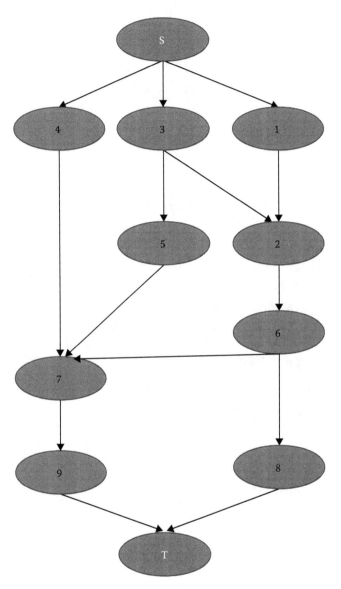

FIGURE 7.2
Bayesian network.

In general, the deadlock conditional probability for the robots is illustrated in Table 7.3.

Then, all other conditional probabilities are computed based on Table 7.2. Here, to compute the reliability of the system, using the posterior failure probabilities and the backward method, two keystone robots 8 and 9 are considered.

TABLE 7.3

Deadlock Conditional Probability

Robot	Conditional Probability	Value
1	$P_{1|2}$	0.93
2	$P_{2|6}$	0.89
3	$P_{3|2}$	0.92
	$P_{3|5}$	0.89
4	$P_{4|7}$	0.90
5	$P_{5|7}$	0.91
6	$P_{6|7}$	0.95
	$P_{6|8}$	0.98
7	$P_{7|9}$	0.99

First, assuming robot 8 as the keystone, the influence is traced using the BN. If robot 8 fails, then robots 1, 2, and 6 are obstructed and the probability of failure and reliability of the system are given:

$$P(\text{sys}) = p(6|8) \cdot p(8) + p(2|6) \cdot p(6) + p(1|2) \cdot p(2).$$

Since robots are similar,

$$P(1) = p(2) = p(3) = \cdots = p(9)$$

$$P(\text{sys}) = (0.185375 \times 0.125) + (0.22125 \times 0.125)$$
$$+ (0.18625 \times 0.125) = 0.07410938$$

$$R(\text{sys}) = 1 - 0.07410938 = 0.92589062.$$

Now, if robot 9 is the keystone, then we have the two paths 4-7 and 3-5-7 influenced by robot 9 failures:

Case I: 4-7-9
$$P(\text{sys}) = p(7|9) \cdot p(9) + p(4|7) \cdot p(7)$$

$$P(\text{sys}) = (0.13375 \times 0.125) + (0.2125 \times 0.125) = 0.04328125$$

$$R(\text{sys}) = 1 - 0.04328125 = 0.95671875$$

Case II: 3-5-7-9

$$P(\text{sys}) = p(7|9) \cdot p(9) + p(5|7) \cdot p(7) + p(3|5) \cdot p(5)$$

$$P(\text{sys}) = (0.13375 \times 0.125) + (0.20375 \times 0.125)$$
$$+ (0.28355 \times 0.125) = 0.07763125$$

$$R(\text{sys}) = 1 - 0.07763125 = 0.92236875$$

According to the results obtained, path 4-7-9 is more reliable for the robots to function in the production system.

7.5 Discussions

In this chapter, a Bayesian network reliability model was proposed to handle a complex system of industrial robots when conditional failure caused by deadlock is present. The model was based on the availability factor of the robots according to the maintenance data for the evaluation of the reliability of the proposed complex systems by using the BN. Applying the Bayesian model, we are able to model real complex systems by including the different links and causalities that can exist between the system components. The Bayesian inference permits the computation of the conditional probability of a given node, which represents the availability of the system in our case, always by taking into account the links and interactions of the system components.

8

Modified Branching Process Reliability Model

KEYWORDS: *complex system reliability, multiple robot, branching process.*

8.1 Introduction

Current design practice is usually to produce a safety system that meets a target level of performance that is deemed acceptable by the regulators. Safety systems are designed to prevent or alleviate the consequences of potentially hazardous events. In many modern industries, the failure of such systems can lead to whole system breakdown. In the reliability analysis of complex systems involving multiple components, it is assumed that the components have different failure rates with certain probabilities. This leads to extensive computational efforts in using the commonly employed generating function (GF) and the recursive algorithm to obtain the reliability of systems consisting of a large number of components. Moreover, when system failure results in fatalities, it is desirable for the system to achieve an optimal rather than an adequate level of performance, given the limitations placed on available resources. This chapter concerns with developing a modified branching process combined with GF to handle the reliability evaluation of a multirobot complex system. The availability of the system is modeled to compute the failure probability of the whole system as a performance measure. The results help decision makers in maintenance departments to analyze critical components of the system in different time periods to prevent system breakdowns.

8.2 Proposed Problem

Consider a complex system including several robots. All robots are functioning at a starting time, but some of them may not be working due to their failures in different time periods. The aim is to find out the probability of system failure due to the breakdowns of significant components of the robots. We model this problem by proposing a modified branching process.

Many real physical systems have (at least approximately) the following structure: They comprise entities of a single type, each of which may produce a random number of further entities of the same type. The entities may be neutrons, individuals in a population, cells, and so on. It is traditional to call them particles and specify their behavior as the one in a branching process. Initially, there is a single particle. Each particle gives rise to a family of particles that replace it in the next generation. Moreover, family sizes are independent and identically distributed random variables (Stirzaker, 2005).

8.3 Branching Formulations

We start in the zeroth time period with one active robot. In the first time period, we shall have 0, 1, 2, 3, ... reliable robots with failure probabilities $p_0, p_1, p_2, ...$ If in the first time period there are k reliable robots, then in the second time period there will be $X_1 + X_2 + ... + X_k$ reliable robots, where $X_1, X_2, ... , X_k$ are independent random variables, each with the common failure probabilities $p_0, p_1, p_2, ...$ This description enables one to construct a tree and a tree measure for any number of time periods.

We now study the probability that the system under consideration stops working (i.e., at some time period there are no reliable robots). Let d_m be the probability that the system stops working in the mth time period. Of course, $d_0 = 0$. In addition, it is clear from the definition that $0 = d_0 \leq d_1 \leq d_2 \leq ... \leq 1$. Hence, d_m converges to a limit d, $0 \leq d \leq 1$, where d is the probability that the system will ultimately stop working. We begin by expressing the value d_m in terms of all possible outcomes in the first time period. If there are j reliable robots in the first time period, then for the system to stop working in the mth time period, each of these lines must stop working in the $(m-1)$th time period. As they proceed independently, this probability is $(d_m - 1)^j$. Therefore,

$$d_m = p_0 + p_1 d_{m-1} + p_2 \left(d_{m-1}\right)^2 + p_3 \left(d_{m-1}\right)^3 + ... \tag{8.1}$$

Let $h(z)$ be the ordinary GF for p_i. Then,

$$h(z) = p_0 + p_1 z + p_2 z^2 + ... \tag{8.2}$$

Using this GF, one can rewrite Equation 8.1 in the following form:

$$d_m = h\left(d_{m-1}\right). \tag{8.3}$$

As $d_m \to d$, we see that using Equation 8.1 the value d that we are looking for satisfies the following equation:

$$d = h(d). \tag{8.4}$$

One solution to Equation 8.4 is always $d=1$, since $1=p_0+p_1+p_2+\ldots$. For the other solutions, first note that the solutions to Equation 8.4 represent intersections of the graphs of $y=z$ and $y=h(z)$, where

$$y = h(z) = p_0 + p_1 z + p_2 z^2 + \ldots \tag{8.5}$$

As $h(0)=p_0$ and the first and second derivatives of $h(z)$ with respect to z are

$$h'(z) = p_1 + 2p_2 z + 3p_3 z^2 + \ldots, \tag{8.6}$$

and

$$h''(z) = 2p_2 + (3)(2)p_3 z + (4)(3)p_4 z^2 + \ldots, \tag{8.7}$$

it turns out that for $z \geq 0$, both derivatives are nonnegative, that is, $h'(z) \geq 0$ and $h''(z) \geq 0$.

8.4 Implementation

It is often desired to know the probability of a robotic system to stop working in a particular time period. In addition, the determination of limits on this probability is beneficial. In this section, some cases are considered to illustrate the application of the proposed method. The aim of these cases is to imply the applicability and flexibility of the proposed method to handle different scenarios involved in computing the reliability of an autonomous robotic system.

This case illustrates the impact of time on a robotic system failure where at the most two independent robots, each with the failure probabilities of $p_0=0.2$, $p_1=0.5$, and $p_2=0.3$, are involved. The proposed approach is applied to this system for 12 time periods. The results that are given in Table 8.1 show that the probability of the system to stop working in time period 12 is approximately 0.6.

As shown in Table 8.1, the failure probability rises sharply as the system operates through the time periods. The result obtained here is useful for the maintenance department to perform preventive maintenance in specific time spans in order to postpone system breakdown in the middle of a production plan. Then, substantial repairs can be planned when the system is off or on holidays.

TABLE 8.1

Probability of the Robot System to Stop Working

Time Periods	Probability of the System to Stop Working
1	0.200000
2	0.312000
3	0.385203
4	0.437116
5	0.475879
6	0.505878
7	0.529713
8	0.549035
9	0.564949
10	0.578225
11	0.589416
12	0.598931

8.5 Discussions

In this chapter, we worked on the reliability evaluation of complex multiple robot systems and proposed a new method. To do this, a modified branching process combined with a GF was presented. Then, system reliability was obtained by calculating the failure probability of the whole system. The results emphasized the flexibility, comprehensive applicability, and effectiveness of the proposed method. For future researches in this area, we recommend the following:

- Calculating the reliability of multistate systems
- Considering repairable components and repair time
- Considering the assumption that some failed components may damage the system

9

Standby Renewal Process
Reliability Model

KEYWORDS: *multiple autonomous robots, reliability evaluation, renewal process, cold standby, environmental factors.*

9.1 Introduction

Providing the required robots is an important issue of availability of a complex robotic production system, which is crucial for system utility improvement. In an advanced production system, the estimation of the required number of robots can be performed through different approaches. One of the realistic autonomous robot estimation methods is based on the system's reliability and the system operating environment. To forecast the required number of robots for the existing production system, in some cases, the assumption of a constant failure rate does not differ much from the assumption of a nonconstant failure rate, and can be made with an acceptable error. In this chapter, we study a renewal model used in the estimation of robots for nonrepairable components. The effectiveness of the proposed integrated reliability evaluation model is worked out in an implementation.

9.2 Statement of the Problem

A multiple autonomous robot production system is proposed so that robots that encounter failures are replaced with a new one. The arrangement of multiple robots and work stations leads to a complex system configuration. Failure of any component of the system results in the whole system to stop functioning, violating on-time product delivery and proper fulfillment of customer demand.

To sum up, most of the applications of the renewal process were focused on finding numerical simplification and optimization, whereas the renewal process can be employed as a useful modeling tool for replacement models. Also, the standby mode was not considered to be integrated with the renewal process. Although, some of the past works considered external factors in renewal models, the effect of environmental factors on the standby renewal process was not investigated. In the cases reported in the literature, complex systems including autonomous robots and a production environment have not been extensively studied.

The environmental conditions in which an equipment is to be operated, such as temperature, humidity, and dust, often have a considerable impact on system reliability (Blischke and Murthy, 2000). Thus, an operating environment should be meticulously maintained when aiming at higher availability and more effective performance. Some of the most significant instances of environmental factors are as follows (Ghodrati and Kumar, 2005):

1. Working environment:
   a. Climatic conditions such as temperature and humidity, in which a system works
   b. Physical environment factors such as dust, smoke, fumes, corrosive agents, and the like
2. User characteristics: Examples are operator skill, education, culture, and language.
3. Operating place or location: This factor refers to workplace settings, such as outside (free) or closed (surrounded) spaces, of the industry that will use the product and/or other area characteristics (such as mines) where the product will be used.
4. Level of application: The system may be intended to have a major purpose, a minor or auxiliary purpose, and even a standby purpose in an operational setup.
5. Work time and period of operation: Planning may call for a product to be in continuous or part-time operation.

Hence, an autonomous robot system design problem in a production environment is considered. The aim is to evaluate the required number of robots considering their failures caused by environmental factors and replacement in a planning horizon.

This problem arises due to expensive equipment and a large amount of investment needed to configure such an advanced automated production system. Thus, accurate planning and effective forecasting are essential to obtain higher productivity. Meanwhile, by evaluating different design scenarios, the estimated number of robots can be obtained so that the availability and the reliability of the system increase.

The renewal process is a useful method for analyzing the number of replacements of autonomous robots in specified time spans for the proposed system. The replaced robots follow a cold standby paradigm in which the replaced component functions properly and guarantees the availability of the system. Then, the defective robots can be handled by maintenance and repair teams to overhaul for future applications. In this process, rapid decision making and action is of importance with regard to demand fulfillment and production due dates.

9.3 Formulation of the Problem

Consider the replacement of an autonomous robot having an average time to failures denoted by $\bar{T}$ and a standard deviation of time to failures denoted by $\sigma(t)$ (so that $\zeta = \dfrac{\sigma(T)}{\bar{T}}$ denotes the coefficient of variation of the time to failures). If the operation time t of the system in which this robot is installed is very long and several replacements need to be made during this period, then the average number of failures $E[N(t)] = M(t)$ will stabilize to the asymptotic value of the *renewal function* as (Gnedenko et al., 1969)

$$N_t = M(t) = E[N(t)] = \frac{t}{\bar{T}} + \frac{\zeta^2 - 1}{2}. \tag{9.1}$$

Equation 9.1 gives the average number of failures in time t. And the corresponding *failure intensity* or *renewal rate function* is given by

$$m(t) = \frac{dM(t)}{dt} = \frac{dE[N(t)]}{dt} = \frac{1}{\bar{T}}. \tag{9.2}$$

The *standard deviation* of the number of failures in time t is given by

$$\sigma[N(t)] = \zeta\sqrt{\frac{t}{\bar{T}}}. \tag{9.3}$$

If time t in the above equations representing a planning horizon is large, then $N(t)$ is approximately normally distributed (based on a central limit theorem) with mean $\overline{N(t)}$. Then the approximated number of autonomous robots N_t needed during this period with a probability of shortage of $1 - p$ is given by

$$N_t = \frac{1}{\bar{T}} + \frac{\zeta^2 - 1}{2} + \zeta\sqrt{\frac{t}{\bar{T}}}\Phi^{-1}(p), \tag{9.4}$$

where $\Phi^{-1}(p)$ is the inverse normal distribution function and is available in probability textbooks.

Assuming the Weibull reliability model (initial failure model) to be a most versatile model for characterizing the life of mechanical parts, and integrating the effect of environmental factors with regard to the multiplicative failure model (i.e., $\theta(t) = \theta(t, z) = \theta_0(t)\omega(z, \alpha)$, where z indicates the environmental factors and α indicates the regression-type parameters (coefficients) for the aforementioned multiplicative model), we have

$$\theta(t) = \frac{\beta_0}{\eta_0} \left(\frac{t}{\eta_0} \right)^{\beta_0 - 1} \exp\left(\sum_{k=1}^{n} \alpha_k z_k \right) = \frac{\beta_0 \cdot t^{\beta_0 - 1}}{\eta_0^{\beta_0}} \exp\left(\sum_{k=1}^{n} \alpha_k z_k \right), \tag{9.5}$$

$$\theta(t) = \frac{\beta_0 \cdot t^{\beta_0 - 1}}{\eta_0^{\beta_0} \cdot \exp\left(-\sum_{k=1}^{n} \alpha_k z_k \right)}, \tag{9.6}$$

$$\theta(t) = \frac{\beta_0 \cdot t^{\beta_0 - 1}}{\left[\eta_0 \cdot \exp\left(-\frac{1}{\beta_0} \sum_{k=1}^{n} \alpha_k z_k \right) \right]^{\beta_0}}. \tag{9.7}$$

This equation indicates the Weibull distribution with the shape parameter and the scale parameter as

$$\beta = \beta_0$$

$$\eta = \eta_0 \left[\exp\left(\sum_{k=1}^{n} \alpha_k z_k \right) \right]^{-\frac{1}{\beta}}. \tag{9.8}$$

Also, it can be concluded that the influencing environmental factors change the scale parameter only, and the shape parameter remains almost unchanged. β_0 and η_0 are the initial shape and scale parameters, respectively, in the Weibull distribution. The coefficient of variation of the time to failures can be calculated based on the shape and scale parameters as follows:

$$\zeta = \frac{\sigma(T)}{\overline{T}}, \tag{9.9}$$

where

$$\overline{T} = \eta \Gamma\left(1 + \frac{1}{\beta} \right), \tag{9.10}$$

$$\sigma(T) = \eta \sqrt{\Gamma\left(1 + \frac{2}{\beta} \right) - \Gamma^2\left(1 + \frac{1}{\beta} \right)}. \tag{9.11}$$

The reliability model obtained by assuming that β_0 is the initial shape parameter and η_0 is the initial scale parameter can be defined as

$$R(t) = 1 - F(t) = \left[1 - \left(1 - \exp\left(-\frac{t}{\eta} \right)^{\beta} \right) \right]$$

$$= \exp\left(-\frac{t}{\eta_0 \left(\exp\left(\sum_{k=1}^{n} \alpha_k z_k \right) \right)^{-\frac{1}{\beta_0}}} \right)^{\beta_0}. \qquad (9.12)$$

Then, considering cold standby for the replacement of autonomous robots, the reliability of the whole system of robots (when the failure rate follows the Weibull probability distribution) with respect to renewal and cold standby is obtained as follows:

$$R_{\text{renewal-cold standby}(n)}(t) = \sum_{j=1}^{n} e^{\left(-\frac{t}{\eta_{0j} \left[\exp\left(\sum_{k=1}^{n} \alpha_k z_k \right) \right]^{-\frac{1}{\beta_j}}} \right)^{\beta_j}}$$

$$\prod_{i} \frac{\beta_i}{\eta_{0i} \left[\exp\left(\sum_{k=1}^{n} \alpha_k z_k \right) \right]^{-\frac{1}{\beta_i}}} \left(\frac{t}{\eta_{0i} \left[\exp\left(\sum_{k=1}^{n} \alpha_k z_k \right) \right]^{-\frac{1}{\beta_i}}} \right)^{\beta_i - 1}. \qquad (9.13)$$

Equation 9.13 gives the reliability of the system of robots when replacements are in the cold standby mode based on the renewal process.

Next, an implementation study is conducted to make use of the developed and proposed reliability and renewal process measures in application.

9.4 Implementation

To analyze different aspects of the proposed multiple robot replacement problem in a complex production system, some cases are considered. This chapter focuses on the influence of the working environment and operator skill as environmental factors on the production system reliability characteristics for optimizing production support (robot provisioning).

TABLE 9.1

Average Number of Robots Obtained with respect to Different Failure Rates

η_0	1,000	2,000	3,000	4,000	5,000	10,000
N_t	6.9	3.4	2.3	1.7	1.3	0.6

The emphasis here is on system design characteristics such as reliability and the required number of autonomous robots based on these influencing environmental factors.

In our case, the aim is to find out and analyze the number of replacements with respect to the effect of environmental factors. The Weibull renewal process model is more appropriate for calculating the total number of available autonomous robots accurately. The calculation process was carried out based on the different values of the initial failure rate (η_0), the shape parameter (β), and the effect of environmental factors (Env. fac. $= \exp\left(\sum_{k=1}^{n} \alpha_k z_k\right)$, which is assumed to be a fixed coefficient). In the implemented calculation process, we used the approximated Weibull renewal model methods for estimating the average required number of autonomous robots in a specified planning horizon (see Table 9.1).

The number of required autonomous robots decreases as the initial failure rate increases. In addition, the slope of the lines is sharp before 3000 hours, and afterward it subsides. We can therefore conclude that for the working period before 3000 hours, it is more beneficial to use the Weibull model, which is more accurate.

9.5 Discussions

Advanced production system availability is dependent not only on the component characteristics, such as reliability and maintainability, but also on the environment in which the component is functioning. Also, the cold standby mode for robots' replacement integrated with the renewal process helps in simplified and combinatorial decision making. Consequently, while calculating the robots' reliability characteristics (failure rate) and predicting the required component (autonomous robot), the robots' operating environment parameters should be taken into account. The remarkable influence of considering and/or ignoring the environmental factors on the forecasting and estimation of the required number of robots is validated by the result of the implementation cases studied. The results implied that the reliability of the whole system is affected if the working environment factors are ignored.

The reliability characteristics of components are usually used for dimensioning components based on engineering specifications. However, a subjective estimate of the effect of environmental factors is considered for the reliability improvement of a system. Future work should be directed more toward obtaining the optimum component support and the average number of components required to prevent stoppages in work. Some suggestions for future research are as follows:

- Developing a redundancy allocation model using the proposed renewal standby process
- Including operational constraints into the model and forming an optimization problem
- Extending the renewal standby model with respect to Markovian state transition for active standby state changes
- Analyzing certain environmental factors to find out optimal conditions for the cost-effective operation of a system
- Investigating the integration of environmental factors in planning the component design for optimizing the component support needs
- Investigating the influence of component environmental factors on the maintenance and support aspects

10

Reliability and Inspection Model

KEYWORDS: *flexible jobshop, autonomous guided vehicle (AGV), manufacturing systems.*

10.1 Introduction

In this chapter, we focus on automated manufacturing systems composed of autonomous guided vehicles (AGVs) as material handling devices and jobshop layout of machines. AGV carries raw materials through shops in batches. In each shop an inspection unit is configured (multistage inspection). Considering the size of the batches and the defect rate, one type of inspection (sampling inspection or full inspection) is carried out. The number of defects in products is a random variable, which specifies the inspection policy. In general, inspection incurs cost on our system. Machines in each shop are in a parallel arrangement. The reliability of machines follows a probability distribution. Shops are in a series arrangement. Hence, a composite (parallel–series) system is resulted. While the reliability of machines decreases, the number of defects in products increases, and as a result it incurs more cost by the system. The aim is to control the number of defects in a way that satisfies both product assurance level, which is a consequence of inspection, and machines reliability lower limit, considering the available operational budgets.

10.2 Problem Description

Advanced automated manufacturing systems are widely used in industries when productivity objectives have to be met. As these systems are often costly, they must be designed to be as efficient as possible. Due to the global competition in the manufacturing field, firms are forced to consider increasing the quality and responsiveness to customization, while decreasing costs. The evolution of flexible manufacturing systems (FMSs) offers great potential for increasing flexibility and changing the basis of competition by ensuring both cost-effective and customized manufacturing at the same time. Some of the important planning problems that need realistic modeling and quicker solutions, especially in automated manufacturing systems, have assumed greater significance in the recent past.

FMSs that are equipped with several computer numerical control (CNC) machines and an AGV-based material handling system are designed and implemented to gain flexibility and efficiency of production. To obtain these benefits, the planning in the FMS decision-making process is critical because the planning decision influences the subsequent decision processes such as scheduling and dispatching. The planning in automated manufacturing systems can be characterized as being online and short term in nature to respond to the frequently changing production order.

In this chapter, we focus on automated manufacturing systems composed of AGVs as material handling devices and jobshop layout of machines. AGV carries raw materials through shops in batches. In each shop an inspection unit is configured (multistage inspection). Fault occurrence in products is a random variable. In general, inspection incurs cost on our system. Machines in each shop are in a parallel arrangement. The reliability of machines follows a probability distribution. Shops are in a series arrangement. Hence, a composite (parallel–series) system is resulted. While the reliability of machines decreases, fault occurrence in products increases and as a result incurs more costs by the system. The aim is to maximize the profit in a way that satisfies both product assurance level, which is a consequence of inspection, and machines reliability lower limit considering the available operational budgets.

Since the number of manufactured products is the decision variable, the proposed costs are coefficients for them. Machines reliability accompanies a cost coefficient. To model the problem, we apply stochastic programming.

10.3 Mathematical Modeling

It is necessary to incorporate reliability into the model to ensure the level of service for each machine in each shop. Reliability is defined as the probability that the system functions until time t. If a machine in a shop breaks down, it can be regarded as a failure. A desired level of reliability can be achieved by limiting the failure probabilities. It is assumed that the reliability of each machine is independent according to the exponential processes. In the following, we discuss the reliability-based model.

$R_j(t)$: The probability of functioning of machine jth in shop ith until time t:

$$R_j(t)_{system} = \begin{cases} \prod_{i=1}^{I}\left(1 - \prod_{j=1}^{J}\left(1 - R_j(t)\right)\right), & \text{machines in parallel} \\ \prod_{i=1}^{I}\left(\prod_{j=1}^{J}R_j(t)\right), & \text{machines in series} \end{cases} \qquad (10.1)$$

In our proposed problem, the machines in each shop are in parallel and the shops are in series, that is, a composite system is configured. Therefore, the reliability of the system is as follows:

$$\prod_{i=1}^{I}\left(1 - \prod_{j=1}^{J}\left(1 - R_j(t)\right)\right) \geq \alpha, \qquad (10.2)$$

where α is the lower bound for a desirable reliability of the system until time t. As previously assumed, the reliability of each machine is independent according to the exponential distribution

$$R_j(t) = e^{\frac{-t}{\theta_j}}, \qquad (10.3)$$

then

$$\prod_{i=1}^{I}\left(1 - \prod_{j=1}^{J}\left(1 - e^{\frac{-t}{\theta_j}}\right)\right) \geq \alpha. \qquad (10.4)$$

It is obvious that to obtain a higher level of reliability, more cost is incurred by the system. Hence, a cost function ($C_j(t)$) is defined to keep machine jth reliable until time t. For the whole system, we have

$$\sum_{j=1}^{J}C_j(t). \qquad (10.5)$$

Thus, the mathematical modeling is as follows.

Assumptions

1. The manufacturing system is jobshop.
2. Material handling is by AGV.
3. There is an inspection station in every shop.

4. Rejected products in the inspection station are returned to the rework cycle.

5. An added value is considered at any stage of production for a product.

6. Reliability for machines is assessable.

7. Reliability follows exponential distribution.

8. Reliability is of composite system.

Indices

i	Index for shop; $i = 1, ..., I$.
$\hat{i}$	Index for shop; $\hat{i} = 1, ..., I$.
j	Index for machine; $j = 1, ..., J$.
$\hat{j}$	Index for machine; $\hat{j} = 1, ..., J$.

Parameters

D	Total demand for the products
B_1	Available budget for reliability expenditures
B_2	Available budget for inspection expenditures
P	Price of a product
C_{rel_j}	Cost of reliability for machine j
CO_{ij}	Operational cost of machine j in shop i
CR_{ij}	Refixturing cost of machine j in shop i
$TR_{ij\hat{i}\hat{j}}$	Transferring cost from machine j in shop i to machine $\hat{j}$ in shop $\hat{i}$
α	Lower bound for reliability
MHC	Material handling cost
VCM	Variable cost of manufacturing
TRE	Total reliability expenditures
dx_{ij}	Differential of decision variable X_{ij}
UM_i	Upper bound for number of machines in shop i
TIC	Total inspection cost
TCT	Total test cost
$TCRW$	Total rework cost
CT_{ij}	Unit test cost in shop i on machine j
CRW_{ij}	Unit rework cost in shop i on machine j
β_{ij}	Fault occurrence in shop i on machine j, which is a stochastic parameter
$E(\cdot)$	Expected value of a parameter
$Var(\cdot)$	Variance of a parameter
Z_p	pth percentile of standard normal distribution

Decision Variables

X_{ij} Number of manufactured products in shop i by machine j

Y_{ij} $\begin{cases} 1 & \text{if machine } j \text{ in shop } i \text{ processes a product} \\ 0 & \text{otherwise} \end{cases}$

M_{ij} $\begin{cases} 1 & \text{if machine } j \text{ is allocated to shop } i \\ 0 & \text{otherwise} \end{cases}$

R_j Reliability for each machine j

Mathematical Model

$$Max\left(p \times \int_0^i X_{ij} \cdot dx_{ij} \right) - MHC - VCM - TRE - TIC \tag{10.6}$$

s.t.

$$TRE = \sum_{j=1}^{J} \sum_{i=1}^{I} \left[\prod_{i=1}^{I} \left(1 - \left(\prod_{j=1}^{J} (1 - R_j) \right) \right) \right] \times X_{ij} \times C_{rel_j}, \tag{10.7}$$

$$MHC = D \times \left(\sum_{j,\hat{j}=1}^{J} \sum_{i,\hat{i}=1}^{I} \left(TR_{ij\hat{i}\hat{j}} \times Y_{ij} \times X_{ij} \right) \right), \tag{10.8}$$

$$VCM = D \times \sum_{j=1}^{J} \sum_{i=1}^{I} \left(\left(CO_{ij} + CR_{ij} \right) \times Y_{ij} \times X_{ij} \right), \tag{10.9}$$

$$TIC = TCT + TCRW, \tag{10.10}$$

$$TCT = \left[\left(\sum_{j=1}^{J} \sum_{i=1}^{I} \left(E(\beta_{ij}) \times X_{ij} \times CT_{ij} \right) \right) + Z_p \cdot \sqrt{\sum_{j=1}^{J} \sum_{i=1}^{I} \left(Cov(\beta_{ij}) \times X_{ij}^2 \times CT_{ij} \right)} \right], \tag{10.11}$$

$$TCRW = \left[\left(\sum_{j=1}^{J}\sum_{i=1}^{I}\left(E\left(\beta_{ij}\right)\times X_{ij}\times CR_{ij}\right)\right) + Z_p \cdot \sqrt{\sum_{j=1}^{J}\sum_{i=1}^{I}\left(Cov\left(\beta_{ij}\right)\times X_{ij}^{2}\times CR_{ij}\right)}\right],$$

$$(10.12)$$

$$\left[\left(\sum_{j=1}^{J}\sum_{i=1}^{I}\left(E\left(\beta_{ij}\right)\times X_{ij}\times CT_{ij}\right)\right) + Z_p \cdot \sqrt{\sum_{j=1}^{J}\sum_{i=1}^{I}\left(Cov\left(\beta_{ij}\right)\times X_{ij}^{2}\times CT_{ij}\right)}\right] +$$

$$\left[\left(\sum_{j=1}^{J}\sum_{i=1}^{I}\left(E\left(\beta_{ij}\right)\times X_{ij}\times CR_{ij}\right)\right) + Z_p \cdot \sqrt{\sum_{j=1}^{J}\sum_{i=1}^{I}\left(Cov\left(\beta_{ij}\right)\times X_{ij}^{2}\times CR_{ij}\right)}\right] \le B_2,$$

$$(10.13)$$

$$\prod_{i=1}^{I}\left(1-\left(\prod_{j=1}^{J}\left(1-R_j\right)\right)\right) \ge \alpha, \qquad (10.14)$$

$$\sum_{j=1}^{J}\sum_{i=1}^{I}\left(\prod_{i=1}^{I}\left(1-\left(\prod_{j=1}^{J}\left(1-R_j\right)\right)\right)\right)\times X_{ij}\times C_{rel_j} \le B_1, \qquad (10.15)$$

$$\sum_{j=1}^{J}\sum_{i=1}^{I}X_{ij} \ge D, \qquad (10.16)$$

$$\sum_{j=1}^{J}Y_{ij} = 1, \quad \forall i, \qquad (10.17)$$

$$\sum_{i=1}^{I}Y_{ij} = X_{ij}, \quad \forall j, \qquad (10.18)$$

$$\sum_{i=1}^{I}Y_{ij} \ge M_{ij}, \quad \forall j, \qquad (10.19)$$

$$\sum_{i=1}^{I}M_{ij} \le 1, \quad \forall j, \qquad (10.20)$$

$$\sum_{j=1}^{J} M_{ij} \le UM_i, \quad \forall i, \tag{10.21}$$

$$Y_{ij}, M_{ij} \in \{0, 1\} \quad \forall i, j, \tag{10.22}$$

$$X_{ij} \in \text{Integer and} \ge 0 \quad \forall i, j, \tag{10.23}$$

$$0 \le R_j \le 1 \quad \forall j. \tag{10.24}$$

In the proposed mathematical model, Equation 10.6 is the objective that is gaining the maximum profit of manufacturing products. The integral implies the added value of the products in any shop. Equation 10.7 presents the total reliability expenditures. Equation 10.8 shows the material handling costs. Equation 10.9 indicates the variable cost of manufacturing. Statement 10.10 indicates that the total inspection cost is an aggregated value of total test cost and total rework cost. Equation 10.11 presents the total test costs. Equation 10.12 shows the total rework costs. Equation 10.13 guarantees that the inspection cost is confined to a specified budget. Equation 10.14 indicates that the lower level for reliability is limited. Equation 10.15 certifies that the available budget for reliability is constrained. Equation 10.16 emphasizes that the amount of production should be equal to or higher than the total demand. Equation 10.17 shows that in any shop, just one machine processes a product. Equation 10.18 certifies that the total number of processed products in any shop on any machine is equal to the total number of manufactured products. Equation 10.19 ensures that a machine has to be allocated to a shop before any operation could be assigned to that machine. Equation 10.20 guarantees that each machine is allocated at the most to one shop. Equation 10.21 certifies that the upper bound for the number of machines in any shop is limited. Equations 10.22 through 10.24 present the kinds of decision variables.

10.4 Implementation

Here, we present an example to illustrate the applicability and effectiveness of our proposed approach. Consider a manufacturing system with five shops and four machines in each shop. The required input data are as follows. The demand to be supplied is 15 units of product, the available budget for reliability is 1500 units of cost, the available budget for inspection is

1200 units of cost, and the lower bound for reliability is 0.05. The following matrices are the remaining inputs:

$$
CO_{ij} = \begin{pmatrix} 5 & 4 & 3 & 7 \\ 4 & 6 & 2 & 5 \\ 3 & 3 & 4 & 3 \\ 2 & 4 & 5 & 4 \\ 4 & 5 & 6 & 6 \end{pmatrix}, \quad
CR_{ij} = \begin{pmatrix} 4 & 2 & 4 & 5 \\ 5 & 4 & 3 & 3 \\ 3 & 1 & 3 & 1 \\ 4 & 2 & 6 & 2 \\ 4 & 3 & 7 & 4 \end{pmatrix}, \quad
CT_{ij} = \begin{pmatrix} 1 & 2 & 2 & 1 \\ 2 & 2 & 1 & 3 \\ 1 & 2 & 1 & 2 \\ 1 & 1 & 2 & 2 \\ 2 & 3 & 1 & 2 \end{pmatrix},
$$

$$
CRW_{ij} = \begin{pmatrix} 3 & 4 & 3 & 4 \\ 5 & 4 & 3 & 2 \\ 3 & 2 & 5 & 3 \\ 2 & 2 & 4 & 4 \\ 3 & 2 & 4 & 5 \end{pmatrix},
$$

$$
\beta_{ij} = \begin{pmatrix} 0.5 & 0.4 & 0.3 & 0.7 \\ 0.4 & 0.6 & 0.2 & 0.5 \\ 0.3 & 0.3 & 0.4 & 0.3 \\ 0.2 & 0.4 & 0.5 & 0.4 \\ 0.4 & 0.5 & 0.6 & 0.6 \end{pmatrix}, \quad
UM_i = \begin{pmatrix} 6 \\ 7 \\ 4 \\ 5 \\ 3 \end{pmatrix}.
$$

Also, the price of a product is set to be 25 and the 95th percentile of standard normal distribution is 1.96. Feeding the input to the optimization software, the following results are obtained for the decision variables:

$$
M_{ij} = \begin{pmatrix} 0 & 1 & 0 & 0 \\ 0 & 0 & 0 & 1 \\ 1 & 0 & 0 & 0 \\ 1 & 0 & 0 & 0 \\ 0 & 0 & 1 & 0 \end{pmatrix}, \quad
Y_{ij} = \begin{pmatrix} 0 & 1 & 0 & 0 \\ 0 & 0 & 0 & 1 \\ 1 & 0 & 0 & 0 \\ 1 & 0 & 0 & 0 \\ 0 & 0 & 1 & 0 \end{pmatrix}, \quad
X_{ij} = \begin{pmatrix} 0 & 3 & 0 & 0 \\ 0 & 0 & 0 & 5 \\ 2 & 0 & 0 & 0 \\ 3 & 0 & 0 & 0 \\ 0 & 0 & 2 & 0 \end{pmatrix},
$$

$$
R_j = \begin{pmatrix} 0.72 & 0.65 & 0.48 & 0.55 \end{pmatrix}.
$$

10.5 Discussions

This work considered an automated manufacturing system composed of AGV as the material handling device and jobshop layout of machines. The AGV carried raw materials through shops in batches. In each shop, an inspection unit was configured. Considering the size of the batches and the defect rate, one type of the inspection (sampling inspection or full inspection) was carried out. In general, inspection incurred cost to our system. Machines in each shop were in a parallel arrangement. The reliability of machines was followed by exponential probability distributions. Shops were in a series arrangement. Hence, a composite (parallel–series) system was resulted. The aim was to control the number of defects in a way that satisfies both product assurance level, which was a consequence of inspection, and machines reliability lower limit, considering the available operational budgets. The efficiency and validity of the proposed model were tested through a numerical example.

References and Further Reading

Abo Al-Kheer, A., El-Hami, A., Kharmanda, M.G., and Morazán, A.M. (2011). Reliability-based design for soil tillage machines. *Journal of Terramechanics* 48(1), 57–64.

Ahmad, S.H. (1988). Simple enumeration of minimal cut sets of acyclic directed graph. *IEEE Transaction on Reliability* 27(5), 484–487.

Akers, B. (1978). Binary decision diagrams. *IEEE Transactions on Computers* 27, 509–516.

AL-Ali, A.A. (1998). Reliability of complex system. *Basra Journal of Science* 16, 112–115.

Aldemir, T. (1987). Computer-assisted Markov failure modeling of process control system. *IEEE Transactions on Reliability* R-36, 133–144.

Alem Tabriz, A., Khorshidvand, B., and Ayough, A. (2015). Modelling age based replacement decisions considering shocks and failure rate. *International Journal of Quality & Reliability Management* 33(1), 107–119.

Andrews, J.D. and Bartlett, L.M. (2005). A branching search approach to safety system design optimization. *Reliability Engineering & System Safety* 87, 23–30.

Apostolakis, G. and Chu, T.L. (1984). Time-dependent accident sequences including human actions. *Nuclear Technology* 64, 115–126.

Avontuur, G.C. (2000). Reliability analysis in mechanical engineering design. PhD thesis, Delft University Press, Delft, the Netherlands.

Avontuur, G.C. and van der Werff, K. (2001). An implementation of reliability analysis in the conceptual design phase of drive trains. *Reliability Engineering & System Safety* 73(2), 155–165.

Baxter, L. (1993). Towards a theory of confidence intervals for system reliability. *Statistics and Probability Letters* 16, 29–38.

Beichelt, F.E. and Fischer, K. (1980). General failure model applied to preventive maintenance policies. *IEEE Transactions on Reliability* R-29, 39–41.

Billinton, R. and Allan, R.N. (1983). *Reliability Evaluation of Engineering Systems: Concepts and Techniques*. Pitman Books Limited, Boston, MA.

Bistouni, F. and Jahanshahi, M. (2014). Analyzing the reliability of shuffle-exchange networks using reliability block diagrams. *Reliability Engineering & System Safety* 132, 97–106.

Blischke, W.R. and Murthy, D.N.P. (1994). *Warranty Cost Analysis*. Marcel Dekker Inc., New York.

Blischke, W.R. and Murthy, D.N.P. (2000). *Reliability—Modeling, Prediction, and Optimization*. John Wiley & Sons, New York.

Blokus, A. (2006). Reliability analysis of large systems with dependent components. *International Journal of Reliability, Quality and Safety Engineering* 13, 1–14.

Bobbio, A. (1988). Use of petri nets for system reliability analysis. In *Ispra Course 'Advanced, Informatic Tools for Safety and Reliability Analysis'*, Ispara, Italy, October 24–28, 1988. Commission of the European Communities, Ispra (VA), Italy, pp. 145–152.

Bobbio, A., Ferraris, C., and Terruggia, R. (2006). New challenges in network reliability analysis. Technical Report, TR-INF-UNIPMN, pp. 1–8.

Boland, P.J., Proschan, F., and Tong, Y.L. (1993). Some recent applications of stochastic inequalities in system reliability theory. In *Advances in Reliability*, A.P. Basu, Ed., North Holland, New York, pp. 29–41.

Bryant, R.E. (1992). Symbolic Boolean manipulation with ordered binary-decision diagrams. *ACM Computing Surveys* 24(3), 293–318.

Carlier, J. and Lucet, C. (1996). A decomposition algorithm for network reliability evaluation. *Discrete Applied Mathematics* 65, 141–156.

Chiacchio, F., Cacioppo, M., D'Urso, D., Manno, G., Trapani, N., and Compagno, L.A. (2013). Weibull-based compositional approach for hierarchical dynamic fault trees. *Reliability Engineering & System Safety* 109, 45–52.

Choi, M.S. and Jun, C.H. (1985). Some variant of polygon-to-chain reductions in evaluating reliability of undirected network. *Microelectronics Reliability* 35(1), 1–11.

Coudert, O. and Madre, J.C. (1992). Implicit and incremental computation of primes and essential primes of Boolean functions. In *Proceedings of the 29th ACM/IEEE Design Automation Conference*, Anaheim, CA, pp. 36–39.

Cox, D.R. (1972). Regression models and life-tables. *Journal of the Royal Statistical Society* B34, 187–220.

Dayal, B. and Singh, J. (1992). Reliability analysis of a system in a fluctuating environment. *Microelectronics Reliability* 32, 601–603.

Der Kiureghian, A. and Song, J. (2008). Multi-scale reliability analysis and updating of complex systems by use of linear programming. *Reliability Engineering & System Safety* 93(2), 288–297.

Dhillon, B.S. and Natesan, J. (1983). Stochastic analysis of outdoor power system in fluctuating environment. *Microelectronics Reliability* 23, 867–881.

Dhillon, B.S. and Rayapati, S.N. (1986). A complex system reliability evaluation method. *Reliability Engineering* 16, 163–177.

Ding, Y. and Lisniaski, A. (2008). Fuzzy universal generating function for multi-state system reliability assessment. *Fuzzy Sets and Systems* 159, 307–324.

Dugan, J.B., Bavuso, S.J., and Boyd, M.A. (1992). Dynamic fault tree for fault tolerant computer systems. *IEEE Transactions on Reliability* 41, 363–376.

Dugan, J.B., Sullivan, K.J., and Coppit, D. (2000). Developing a low cost high quality software tool for dynamic fault tree analysis. *IEEE Transactions on Reliability* 49, 49–59.

Ebeling, C.E. (1997). *An Introduction to Reliability and Maintainability Engineering*. McGraw-Hill, Boston, MA, pp. 23–32.

Enevoldsen, I. and Sørensen, J.D. (1994). Reliability-based optimization in structural engineering. *Structural Safety* 15(3), 169–196.

Felix de Oliveira, C.C., Tadeu Cristino, C., and Alves Firmino, P.R. (2016). In the kernel of modelling repairable systems: A goodness of fit test for Weibull-based generalized renewal processes. *Journal of Cleaner Production* 133, 358–367.

Fiorenzo, M. (2008). Automation and robotic in construction, new challenge for old and new industrialized countries. *Automation in Construction* 17(2), pp. 109–110.

Gadani, J.P. (1981). System effectiveness evaluation using star and delta transformations. *IEEE Transaction on Reliability* 30(1), 43–47.

Ge, D., Lin, M., Yang, Y., Zhang, R., and Chou, Q. (2015). Quantitative analysis of dynamic fault trees using improved sequential binary decision diagrams. *Reliability Engineering & System Safety* 142, 289–299.

Gertsbakh, I.B. (1989). *Statistical Reliability Theory*. Marcel Dekker, New York.

Ghodrati, B. and Kumar, U. (2005). Reliability and operating environment based spare parts estimation approach: A case study in Kiruna Mine, Sweden. *Journal of Quality in Maintenance Engineering* 11(2), 169–184.

Gnedenko, B. and Ushakov, I. (1995). *Probabilistic Reliability Engineering*. Wiley, New York.

Gnedenko, B.V., Belyayev, Y.K., and Solovyev, A.D. (1969). *Mathematical Methods of Reliability*. Academic Press, New York.

Goel, P. and Singh, J. (1995). Reliability analysis of a standby complex system having imperfect switch over device. *Microelectronics Reliability* 35, 285–288.

Govil, A.K. (1983). *Reliability Engineering*. McGraw Hill, New Delhi.

Guo, J. and Wilson, A.G. (2013). Bayesian methods for estimating system reliability using heterogeneous multilevel information. *Technometrics* 55, 461–472.

Haftka, R.T. and Gürdal, Z. (1992). *Elements of Structural Optimization*, 3rd edn. Kluwer, New York.

Hao, J., Zhang, L., and Wei, L. (2014). Reliability analysis based on improved dynamic fault tree. In *Proceedings of the Sixth World Congress on Engineering Asset Management 2011*, J. Lee, J. Ni, J. Sarangapani, J. Mathew. Eds., Springer, London, pp. 283–299.

Hariri, S. and Raghavendra, C.S. (1987). SYREL: A symbolic reliability algorithm based on path and cut set methods. *IEEE Transaction on Computers* 36(10), 1224–1232.

Hassan, M. and Aldemir, T. (1990). A data base oriented dynamic methodology for the failure analysis of closed loop control systems in process plants. *Reliability Engineering & System Safety* 27, 275–322.

Hickman, J.W. (1983). *PRA Procedures Guide: A Guide to the Performance of Probabilistic Risk Assessments for Nuclear Power Plants*. NUREG/CR-2300, Vol. 1. Office of Nuclear Regulatory Research, Washington, DC.

Hill, S.D. and Spall, J.C. (1998). Inequality-based reliability estimates for complex systems. In *Proceedings of the American Control Conference*, Philadelphia, PA, June 24–26, pp. 1177–1179.

Hoyland, A. and Rausand, M. (2004). *System Reliability Theory: Models, Statistical Methods, and Applications*, 2nd edn. John Wiley & Sons, Inc., Hoboken, NJ.

Hryniewicz, O. (2006). An evaluation of the reliability of complex systems using shadowed sets and fuzzy lifetime data. *International Journal of Automation and Computing* 3, 145–150.

Huang, H.Z., Xu, H.W., and Zu, X. (2010). Petri net-based coordination component for collaborative design. *Concurrent Engineering—Research and Applications* 18, 199–205.

Hwang, C.L., Tillman, F.A., and Lee, M.H. (1981). System-reliability evaluation techniques for complex/large systems—A review. *IEEE Transactions on Reliability* R-30, 416–423.

Imai, H., Sekine, K., and Imai, K. (1999). Computational investigations of all terminal network reliability via BDDs. *IEICE Transactions on Fundamentals* 82(5), 714–721.

Jardine, A.K.S. (1998). *Maintenance, Replacement and Reliability*. Preney Print and Litho Inc., Toronto, Ontario, Canada.

Jeong, K.S., Chang, S.H., and Kim, T.W. (1987). Development of the dynamic fault tree using Markovian process and supercomponents. *Reliability Engineering & System Safety* 19, 137–160.

Juran, J. and Gryna, F. (1988). *Quality Control Handbook*, 4th edn. McGraw-Hill, New York.

Kallen, M.J. (2011). Modelling imperfect maintenance and the reliability of complex systems using superposed renewal processes. *Reliability Engineering & System Safety* 96(6), 636–641.

Kapur, K.C. and Lamberson, L.R. (1977). *Reliability in Engineering Design*. John Wiley & Sons, New York, pp. 8–30.

Korayem, M.H. and Iravani, A. (2008). Improvement of 3P and 6R mechanical robots reliability and quality applying FMEA and QFD approaches. *Robotics and Computer-Integrated Manufacturing* 24, 472–487.

Kumamoto, H. and Henley, E.J. (2000). *Probabilistic Risk Assessment and Management for Engineers and Scientists*, 2nd edn. Wiley-IEEE Press, Piscataway, NJ.

Kumar, D. and Klefsjö, B. (1994). Proportional hazards model: A review. *Reliability Engineering & System Safety* 44(2), 177–188.

Kumar, D., Singh, J., and Singh, I.P. (1988). Reliability analysis of the feeding system in chapter industry. *Microelectronics Reliability* 28, 213–215.

Kumar, U.D., Crocker, J., Knezevic, J., and El-Haram, M. (2000). *Reliability, Maintenance and Logistic Support: A Life Cycle Approach.* Kluwer Academic Publishers, Boston, MA.

Kuo, W. and Wan, R. (2007). Recent advances in optimal reliability allocation. *IEEE Transactions on System Man, Cybernetics Part A: Systems and Humans* 37, 143–156.

Kuo, W. and Zuo, M.J. (2002). *Optimal Reliability Modeling, Principles and Applications,* 1st edn. John Wiley & Sons, Inc., Hoboken, NJ.

Kuo, W. and Zuo, M. (2003). *Optimal Reliability Modeling: Principles and Applications.* John Wiley & Sons, Hoboken, NJ.

Kuschel, N. and Rackwitz, R. (1997). Two basic problems in reliability-based structural optimization. *Mathematical Methods of Operations Research* 46(3), 309–333.

Lawless, J.F. (1983). Statistical methods in reliability (with discussion). *Technometrics* 25(4), 305–335.

Leroy, A. (1989). Economic study of the need to keep an emergency pipeline repair system on stand-by. *The SRS Quarterly Digest,* Vol. 1, 10–14.

Levitin, G. and Lisnianski, A. (1998a). Joint redundancy and maintenance optimization for multi-state series–parallel systems. *Reliability Engineering & System Safety* 64, 33–42.

Levitin, G. and Lisnianski, A. (1998b). Structure optimization of power system with bridge topology. *Electrical Power Systems Research* 45, 201–208.

Levitin, G. and Lisnianski, A. (2013). *Multi-State System Reliability: Assessment, Optimization and Applications.* World Scientific Publishing, Singapore.

Levitin, G., Lisnianski, A., Ben-Haim, H., and Elmakis, D. (1998). Redundancy optimization for series–parallel multi-state systems. *IEEE Transaction on Reliability* 47, 165–172.

Li, W. and Zuo, M. (2008). Reliability evaluation of multi-state weighted k-out-of-n systems. *Reliability Engineering & System Safety* 93, 160–167.

Li, Y.F., Huang, H.Z., Liu, Y., Xiao, N.C., and Li, H. (2012). A new fault tree analysis method: Fuzzy dynamic fault tree analysis. *Eksploat Niezawodn* 14, 208–214.

Li, Y.F., Mi, J., Huang, H.Z., Xiao, N.C., and Zhu, S.P. (2013). System reliability and assessment for solar array drive assembly based on Bayesian networks. *Eksploat Niezawodn* 15, 117–122.

Lin, J., Lnordenvaad, M.L., and Zhu, H. (2011). Bayesian survival analysis in reliability for complex system with a cure fraction. *International Journal of Performability Engineering* 7, 109–120.

Lindhe, A., Norberg, T., and Rosén, L. (2012). Approximate dynamic fault tree calculations for modelling water supply risks. *Reliability Engineering & System Safety* 106, 61–71.

Lisnianski, A. and Levitin, G. (2003). *Multi-state System Reliability Assessment, Optimization, Applications.* World Scientific, Singapore.

Lisnianski, A., Levitin, G., and Ben-Haim, H. (2000). Structure optimization of multi-state system with time redundancy. *Reliability Engineering & System Safety* 67, 103–112.

Lisnianski, A., Levitin, G., Ben-Haim, H., and Elmakis, D. (1996). Power system structure optimization subject to reliability constraints. *Electrical Power Systems Research* 39, 145–152.

Lucet, C. and Manouvrier, J.F. (1999). Exact methods to compute network reliability. In *Statistical and Probabilistic Models in Reliability.* D.C. Ionescu and N. Limnios, Eds. Birkhauser, Boston, MA, pp. 279–294.

MacLeod, E.N. and Chiarella, M. (1993). Navigation and control breakthrough for automated mobility. In *Proceedings of the SPIE Mobile Robotics VIII*, Boston, MA, September 9–10, Vol. 2058, pp. 57–68.

Mahajan, P. and Singh, J. (1999). Reliability of utensils manufacturing plant—A case study. *Opsearch* 36, 260–271.

Mak, T.M. (2007). Infant mortality—The lesser known reliability issue. In *Proceedings of the 13th IEEE International On-Line Testing Symposium (IOLTS 2007)*, Crete, pp. 122–123.

Maranzano, C.J. and Spall, J.C. (2010). Robust test design for reliability estimation with modeling error when combining full system and subsystem tests. In *Proceedings of the American Control Conference*, Baltimore, MD, pp. 3741–3746, Number ThB17.6.

Matsuoka, T. and Kobayashi, M. (1988). GO-FLOW: A new reliability analysis methodology. *Nuclear Science & Engineering* 9(8), 64–78.

Merle, G., Roussel, J.M., Lesage, J., and Bobbio, A. (2010). Probabilistic algebraic analysis of fault trees with priority dynamic gates and repeated events. *IEEE Transactions on Reliability* 59, 250–261.

Mettas, A. (2000). Reliability allocation and optimization for complex systems. In *Annual Reliability and Maintainability Symposium*, Los Angeles, CA, pp. 216–221.

Mi, J., Li, Y.F., Yang, Y.J., Peng, W., and Huang, H.Z. (2016). Reliability assessment of complex electro mechanical systems under epistemic uncertainty. *Reliability Engineering & System Safety* 152, 1–15.

Mo, Y. (2014). A multiple valued decision diagram based approach to solve dynamic fault trees. *IEEE Transactions on Reliability* 63, 81–93.

Nishijima, K. (2007). *Optimal Reliability of Components of Complex Systems Using Hierarchical System Models.* Institute of Structural Engineering, ETH Zurich, Switzerland.

O'Connor, P.D.T. (2002). *Practical Reliability Engineering*, 4th edn. John Wiley & Sons, New York.

O'Halloran, B.M., Hoyle, C., Stone, R.B., and Tumer, I.Y. (2012). The early design reliability prediction method. In *Proceedings of the ASME International Mechanical Engineering Congress and Exposition (IMECE '12)*, Houston, TX, pp. 1765–1776.

Ouzineb, M., Nourelfath, M., and Gendreau, M. (2008). Tabu search for the redundancy allocation problem of homogenous series–parallel multi-state systems. *Reliability Engineering & System Safety* 93, 1257–1272.

Pedrycz, W. (1998). Shadowed sets: Representing and processing fuzzy sets. *IEEE Transactions on Systems, Man, and Cybernetics—Part B: Cybernetics* 28, 103–109.

Pourkarim Guilani, P., Sharifi, M., Niaki, S.T.A., and Zaretalab, A. (2014). Reliability evaluation of non-reparable three-state systems using Markov model and its comparison with the UGF and the recursive methods. *Reliability Engineering & System Safety* 129, 29–35.

Provan, J.S. (1986). The complexity of reliability computations on planar and acyclic graphs. *SIAM Journal of Computing* 15(3), 694–702.

Qureshi, Z.H. (2007). A review of accident modelling approaches for complex socio-technical systems. In *Proceedings of the 12th Australian Workshop on Safety Related Programmable Systems (SCS'07)*, Adelaide, Australia. Conferences in Research and Practice in Information Technology, Vol. 86. Tony Cant, Ed., Australian Computer Society, Adelaide, Australia, pp. 47–59.

Rajabalinejad, M. (2010). Bayesian Monte Carlo method. *Reliability Engineering & System Safety* 95, 1050–1060.

Rao, K.D., Gopika, V., Rao, V.S., Kushwaha, H.S., Verma, A.K., and Srividya, A. (2009). Dynamic fault tree analysis using Monte Carlo simulation in probabilistic safety assessment. *Reliability Engineering & System Safety* 94, 872–883.

Rao, M.M.S. and Naikan, V.N.A. (2016). A Markov system dynamics approach for repairable systems reliability modeling. *International Journal of Reliability, Quality and Safety Engineering* 23, 1650004.

Rauzy, A. (1993). New algorithms for fault tolerant trees analysis. *Reliability Engineering & System Safety* 5(59), 203–211.

Rauzy, A. (2003). A new methodology to handle Boolean models with loops. *IEEE Transactions on Reliability* 52(1), 96–105.

Rigdon, S.E. and Basu, A.P. (2000). *Statistical Methods for the Reliability of Repairable Systems*. John Wiley & Sons, New York.

Ross, S.M. (1970). *Applied Probability Models with Optimisation Applications*. Holden-Day, San Francisco, CA.

Ross, S.M. (1997). *Introduction to Probability Models*, 6th edn. Academic Press, San Diego, CA.

Ross, S.M. (2010). *Introductory Statistics*, 3rd edn. Academic Press, San Diego, CA.

Royset, J.O., Der Kiureghian, A., and Polak, E. (2001). Reliability-based optimal structural design by the decoupling approach. *Reliability Engineering & System Safety* 73(3), 213–221.

Saleh, J.H. and Marais, K. (2006). Highlights from the early (and pre-) history of reliability engineering. *Reliability Engineering & System Safety* 91(2), 249–256.

Satyanarayana, A. and Chang, M.K. (1983). Network reliability and the factoring theorem. *Networks* 13, 107–120.

Shojaeifar, A., Fazlollahtabar, H., and Mahdavi, I. (2016). Decomposition versus minimal path and cuts methods for reliability evaluation of an advanced robotic production system. *Journal of Automation, Mobile Robotics & Intelligent Systems* 10(3), 52–57.

Silvestri, T. (2014). Complex system reliability: A graph theory approach. *The Mathematica Journal* 16, 1–19.

Singh, J. (1989). A warm stand by redundant system with common cause failures. *Reliability Engineering & System Safety* 26, 135–141.

Singhal, M., Chauhan, R.K., and Sharma, G. (2010). An alternate approach to compute the reliability of a computer communication network using binary decision diagrams. *Communications in Computer and Information Science* 94(4), 160–170.

Singhal, M., Chauhan, R.K., and Sharma, G. (2011). A new approach for finding the various optimal variable ordering to generate the binary decision diagrams (BDD) of a computer communication network. *International Journal of Computer Applications* 31(3), 1–8.

Singhal, M., Chauhan, R.K., and Sharma, G. (2012). Binary decision diagrams and its variable ordering for disjoint network. *International Journal of Advanced Networking and Applications* 3(6), 1430–1437.

Siu, N. and Acosta, C. (1991). Dynamic event tree analysis: An application to SGTR. In *Proceedings of the International Conference Probabilistic Safety Assessment and Management (PSAM)*, G.E. Apostolakis, Ed. Elsevier Science Publishers, London, U.K., pp. 539–541.

Smidts, C. (1990). Simulation des srquences industrielles accidentelles prenant en compte le facteur humaine. Application au domaine des centrales nucleaires. PhD Thesis, Universitd Libre de Bruxelles, Bruxelles, France.

Soleimani, M. and Pourgol-Mohammad, M. (2014). Design for reliability of complex system with limited failure data: Case study of a horizontal drilling equipment. In *Proceedings of the Probabilistic Safety Assessment and Management PSAM, 12*, Honolulu, HI, pp. 21–28.

Soleimani, M., Pourgol-Mohammad, M., Rostami, A., and Ghanbari, A. (2014). Design for reliability of complex system: Case study of horizontal drilling equipment with limited failure data. *Journal of Quality and Reliability Engineering* 1, 13pp. Article ID 524742.

Spall, J.C. (2010). Convergence analysis for maximum likelihood-based reliability estimation from subsystem and full system tests. In *Proceedings of the 49th IEEE Conference on Decision and Control*, Atlanta, GA, pp. 2017–2022.

Srinath, L.S. (1985). *Concepts in Reliability Engineering*, 2nd edn. Affiliated East–West Press, New Delhi, India.

Stamatelatos, M., Vesely, W., Dugan, J.B., Fragola, J., Minarick, J., and Railsback, J. (2002). *Fault Tree Handbook with Aerospace Applications*. NASA Office of Safety and Mission Assurance, Washington, DC.

Stirzaker, D. (2005). *Stochastic Processes and Models*. Oxford University Press, New York.

Tanwar, M., Rai, R.N., and Bolia, N. (2014). Imperfect repair modeling using Kijima type generalized renewal process. *Reliability Engineering & System Safety* 124, 24–31.

Theologou, O. and Carlier, J. (1991). Factoring and reductions for networks with imperfect vertices. *IEEE Transaction on Reliability* 40, 210–217.

Tian, Z., Levitin, G., and Zuo, M. (2009). A joint reliability–redundancy optimization approach for multi-state series–parallel systems. *Reliability Engineering & Systems Safety* 94, 1568–1576.

Torres-Toledano, J.G. and Sucar, L.E. (1998). Bayesian networks for reliability analysis of complex systems. In *Progress in Artificial Intelligence—IBERAMIA 98*. Springer, London, U.K., pp. 195–206.

Uematsu, K. and Nishida, T. (1987). The branching nonhomogeneous Poisson process and its application to a replacement model. *Microelectronics Reliability* 27, 685–691.

Ushakov, I. (1986). Universal generating function. *Soviet Journal Computer Systems Science* 24, 118–129.

Ushakov, I. (2000). The method of generalized generating sequences. *European Journal of Operational Research* 125, 316–323.

Utkin, L.V. and Kozine, I.O. (2001). Computing the reliability of complex systems. In *Proceedings of the 2nd International Symposium on Imprecise Probabilities and Their Applications*, Ithaca, NY, pp. 1–8.

van der Weide, J.A.M. and Pandey, M.D. (2015). A stochastic alternating renewal process model for unavailability analysis of standby safety equipment. *Reliability Engineering & System Safety* 139, 97–104.

Vesely, W.E. and Goldberg, F.F. (1977). Time dependent unavailability analysis of nuclear safety systems. *IEEE Transactions on Reliability* R-264, 257–260.

Wang, Z.-M. and Yang, J.-G. (2012). Numerical method for Weibull generalized renewal process and its applications in reliability analysis of NC machine tools. *Computers & Industrial Engineering* 63(4), 1128–1134.

Williams, R.L. and Gateley, W.Y. (1978). GO methodology: Overview. EPRI NP-765, Electric Power Research Institute, Palo Alto, CA.

Wong, K. (1981). Unified field (failure) theory: Demise of the Bathtub curve. In *Proceedings of Annual RAMS*, Reno, NV, pp. 402–408.

Wood, R.K. (1985). A factoring algorithm using polygon-to-chain reductions for computing K-terminal network reliability. *Networks* 15, 173–190.

Yevkin, O. and Krivtsov, V. (2012). An approximate solution to the G–renewal equation with an underlying Weibull distribution. *IEEE Transactions on Reliability* 61(1), 68–73.

Zio, E. and Pedroni, N. (2009). Building confidence in the reliability assessment of thermal-hydraulic passive systems. *Reliability Engineering & System Safety* 94, 268–281.

Index